CABI CONCISE

MEDICINAL AND AROMATIC PLANTS

Aromatic plants and spices are probably the most underestimated food worldwide. Although they have constantly accompanied humans in the course of evolution (co-evolution), they are only rarely found in official dietary recommendations. Yet they are not only aromatic, but extremely effective and can be used for everything from increasing well-being to treating various illnesses. They are, so to speak, the 'tastiest medicine in the world'. The essential oils contained in most spices, for example, not only give the spice its specific flavour, but also have various antibacterial, antiviral and anti-inflammatory properties.

There is a very long tradition of using herbs and spices for health. In traditional medicinal systems such as Ayurvedic, Chinese, European Monastic and Persian medicine, spices have been used to treat various diseases for more than 3000 years. Today, traditional medicinal systems are a topic of global importance. In many countries, a large proportion of the population relies heavily on traditional practitioners and medicinal plants to meet the need for basic medical care. This is because modern synthetic preparations are often only available to a limited extent and are usually very expensive. Medicinal and aromatic plants (MAPs), or herbal medicines, on the other hand, have often retained their popularity for historical and cultural reasons.

In recent decades, increasing attention has been paid to the phytochemical and pharmacological study of traditional medicinal plants in order to discover and develop new, safe and affordable medicines. These medicinal plants are not only derived from the well-known traditional medical systems but also refer to the experiential knowledge of indigenous people and tribes in South America and Africa. For example, the root of devil's claw (*Harpagophytum procumbens*) was used extensively for healing purposes by the first inhabitants of South Africa.

Since human olfactory cells have direct access to the oldest regions of our brain, such as the limbic system and the hippocampus, where emotions, feelings and, above all, our memories are stored, scents have a direct effect on our well-being. It therefore seems worthwhile to consider how traditional knowledge about the health benefits of various medicinal and aromatic plants can be re-evaluated using the techniques available today, and how the results can be utilized for the pharmaceutical and food sectors.

This series focuses on traditional medicine systems and covers different medicinal plant species or products. Each book will largely follow the same content structure, and will cover the history, efficacy, active ingredients, occurrence, genetics/breeding, cultivation and economic importance.

As the market demand for natural ingredients is increasing, this series aims to provide succinct and readable overviews of key medicinal and aromatic plant species used around the world. We welcome new proposals and suggestions.

Series Editor
Hartwig Schulz
hs.consulting.map@t-online.de

Titles Available

1. Essential Oils Unveiled: Complex Compositions for Food, Cosmetics, and Medicine
   Marek Bunse, Hartwig Schulz *et al.*
2. Saffron: Bridging Traditional and Modern Applications in Nutrition and Medicine
   Hassan Rezadoost, Hassan Esmaeili, Edris Mahdavi-Fikjvar and Hartwig Schulz

Titles in preparation

Cardamom: Advances in Breeding, Production and Quality
R. Sivaranjani and H.J. Akshitha

Saffron

Bridging Traditional and Modern Applications in Nutrition and Medicine

---

Hassan Rezadoost

Medicinal Plants and Drugs Research Institute, Shahid Beheshti University, Tehran, Iran

Hassan Esmaeili

Medicinal Plants and Drugs Research Institute, Shahid Beheshti University, Tehran, Iran

Edris Mahdavi-Fikjvar

Medical Biotechnology Research Center, School of Paramedicine, Guilan University of Medical Sciences, Rasht, Iran

Hartwig Schulz

Consulting and Project Managing for Medicinal and Aromatic Plants, Stahnsdorf, Germany

CABI

CABI is a trading name of CAB International

CABI
Nosworthy Way
Wallingford
Oxfordshire OX10 8DE
UK

Tel: +44 (0)1491 832111
E-mail: info@cabi.org
Website: www.cabi.org

CABI
200 Portland Street
Boston
MA 02114
USA

T: +1 (617)682-9015
E-mail: cabi-nao@cabi.org

A catalogue record for this book is available from the British Library, London, UK.

ISBN-13: 9781836995135 (hardback)
9781836995142 (paperback)
9781836995159 (ePDF)
9781836995166 (ePub)

DOI: 10.1079/9781836995166.0000

Commissioning Editor: Rebecca Stubbs
Editorial Assistant: Theresa Regueira
Production Editor: Theresa Regueira

Typeset by Exeter Premedia Services Pvt Ltd, Chennai, India
Printed and bound by CPI Group (UK) Ltd, Croydon, CR0 4YY

Contents

Preface

For thousands of years, saffron has occupied a unique place in history, culture, medicine and science. Its bright red stigmas are prized not only for their culinary excellence, but also for their deep-rooted medicinal, cultural and spiritual significance.

This book provides in-depth and up-to-date information on saffron, highlighting its historical, botanical, chemical and industrial aspects. Particular attention is paid to the latest scientific findings and forward-looking perspectives. It is aimed especially at readers who already have a basic knowledge of the subject and wish to deepen their understanding of this extraordinary plant.

Saffron has always been a symbol of luxury, spirituality and healing. Its use dates back several millennia and is particularly deeply rooted in the Middle East. Iran, which is still the world's largest producer today, played a central role in this: even in ancient Persia, saffron was valued both in religious ceremonies and as a medicinal plant.

In the modern world, saffron is one of the most valuable spices due to its unique aroma, taste and intense colouring power, as well as its many health benefits. Cultivation and harvesting are still largely done by hand using traditional methods, which contributes significantly to its high market value. At the same time, innovative approaches are finding their way into saffron cultivation, including optimized irrigation systems, improved soil management and targeted measures to combat plant diseases.

With the increasing demand for organic saffron, sustainable cultivation methods are increasingly becoming the focus of research. Post-harvest processes such as drying and storage are also the subject of intensive research and development work, as they play a crucial role in preserving the bioactive ingredients and sensory properties.

The exceptional properties of saffron are mainly due to its bioactive compounds, in particular crocin, picrocrocin and safranal. A thorough understanding of its chemical composition is therefore essential for quality control, authenticity testing and scientific research into its health-promoting effects.

Hartwig Schulz
(hs.consulting.map@t-online.de)

Acknowledgements

I would like to express my deepest gratitude to Dr Fatemeh Mirzajani, Associate Professor at Faculty of Life Sciences and Biotechnology, Shahid Beheshti University, for her continuous support throughout the writing process of this book.

I also wish to thank Dr Aioub Moradi, Professor at the Agricultural Research, Education and Extension Organization of Guilan Province, for his valuable guidance and consultation on saffron botany.

Hassan Rezadoost

Introduction: Historical and Cultural Background, Significance in the World Today

1

Hassan Rezadoost and Hartwig Schulz

1.1 Introduction and Historical Background

Saffron, derived from the dried purple stigmas of *Crocus sativus* (Fig. 1.1), is one of humanity's most valuable and symbolically rich plant-based drug. For over three millennia, it has occupied a unique position at the intersection of agriculture, trade, medicine, cuisine and culture. Saffron's extraordinary value, often exceeding that of gold by weight, has shaped economies and civilizations, making it far more than a simple spice. It is a botanical artefact of ancient societies, a motif in art and ritual, and a modern agricultural product that continues to influence global markets and scientific research.

The history of saffron spans the ancient civilizations of the Mediterranean, the Middle East and South Asia. Archaeological evidence suggests its use as early as 1600 BCE in Minoan Crete, where frescoes depict saffron gatherers harvesting the delicate flowers (Kazemi-Shahandashti *et al.*, 2022; Wikimedia Commons, 2026). Ancient Persia (modern-day Iran) is considered one of the earliest centres of saffron cultivation, where it became intertwined with royal ceremonies, textile dyeing and early medicinal practices. Persian caravans helped spread saffron eastward along the Silk Road to India and China, while Phoenician and Greek traders carried it westward into Europe (López-Sampson and Page, 2018; Bahrami, 2020).

Classical authors such as Homer, Pliny the Elder and Dioscorides referenced saffron as a potent medicinal and aromatic substance (López-Sampson and Page, 2018). In ancient Egypt, Cleopatra reportedly used saffron in her baths for its cosmetic and aphrodisiac qualities (Dalby, 2000). Throughout the medieval period, saffron became central to European trade, particularly in Italy, Spain and later in the Netherlands, where it was used to treat diseases, dye fabrics and refine haute cuisine. Its value was so high that fraudulent

Fig. 1.1. *Crocus sativus* (photograph by Edris Mahdavi).

adulteration was punished with severe penalties, even death in some regions (Davis, 2011).

1.2 Cultural Significance

Across cultures, saffron has symbolized luxury, wisdom, purity and spiritual illumination (Wright, 1999). In South Asia, saffron-coloured robes are worn by monks and ascetics, symbolizing renunciation and enlightenment (Brockington, 2019). In Persian poetry, saffron is used to symbolize joy, celebrations and the arrival of spring, as well as sholezard (saffron and rice), a culinary tradition. In Mediterranean culinary traditions – such as Spanish paella, Italian risotto and French bouillabaisse – saffron gives dishes their golden hue and distinctive aroma, making it an integral part of the national identity (Kuster, 2003) .

Saffron continues to play a role in religious and cultural rituals. It is used in Hindu ceremonies, Persian New Year celebrations (Nowruz) and traditional Middle Eastern perfumery. Historically, its use in textiles and dyes was a sign of high social status and artistic craftsmanship. Even today, saffron remains a symbol of refinement and cultural heritage, representing continuity with the ancient world.

1.3 Significance in the Modern World

In the 21st century, saffron maintains its prestige while adapting to new scientific, economic and global contexts. Iran remains the dominant producer, supplying more than 85% of the world's saffron, followed by India (particularly Kashmir), Spain, Afghanistan, Greece and Morocco. Despite modern agricultural technology, saffron production is still labour-intensive – each flower contains only three delicate stigmas, requiring thousands of blossoms to produce a single ounce. This scarcity sustains its high value and influence in agricultural economies.

Modern research has renewed interest in saffron's pharmacological properties. Studies suggest potential benefits for mood disorders, cognitive function, cardiovascular health and antioxidant activity, making saffron a promising natural therapeutic agent. In gastronomy, saffron remains a global luxury ingredient, prized by chefs for its ability to elevate dishes with nuanced flavour and colour.

Economically, saffron has become a catalyst for rural development and women's employment, particularly in regions where alternative crops are unprofitable. In Afghanistan, for instance, saffron cultivation is promoted as a sustainable alternative to opium production. Meanwhile, climate change, fraudulent substitution and supply chain complexity pose current challenges to saffron's authenticity and sustainability.

1.3.1 Global economic significance

Iran has long been known as a benchmark for saffron cultivation, even today. At present, Iran alone now provides between 85 and 90% of global saffron production, which is due to a number of special climatic, geographical and endemic factors that contribute to this successful situation (FAO, 2022). The Iranian provinces of Khorasan Razavi and South Khorasan are known as the centre of saffron cultivation. They have ideal air-dry soils with rich mineral properties, and a dedicated workforce with many years of experience with saffron cultivation (Behzadi and Shahsavand, 2009). Once, in the past, Persia led all other lands in cultivated saffron, which it obtained from 'a plant with a yellow root, a middle of coralloid colour, and a reddish stalk'. This refers to saffron, a plant known for over 2700 years, starting from Persian civilization (Kabiri *et al.*, 2017). Iranian saffron has a long history of being used in food, traditional Iranian medicine, magical rituals and in Iranian symbols, which all highlight this plant that 'originates in Iraq, but has been naturalized in Mesopotamia since ancient Persian times' (Katouzian, 2009). Saffron has also played an important role in maintaining a livelihood for hundreds of thousands of people in Iran, as a major crop for Iranian farmers growing it in their backyards. Saffron in Iran has been a major branch of the Iranian economy in recent decades. More particularly, Iranian saffron, as a major product of Iran, has been exported to other countries like Spain, the United Arab Emirates (UAE) and even China, where it is further processed in large quantities in order to generate additional profits (Freedman, 2008; Mohammadi and Reed, 2020).

Afghanistan has emerged in recent decades as one of the most promising new regions for saffron cultivation, driven largely by efforts to provide farmers with a viable and legal alternative to opium poppy production

(United Nations Office on Drugs and Crime [UNODC], 2021). For more than 40 years, Afghanistan has been the world's leading producer of illicit opium, a crop that has fuelled insecurity, corruption and conflict. International development agencies, Afghan institutions and local communities have increasingly turned to saffron as a substitute capable of generating income comparable to, or even superior to, that of other crops while promoting social and economic stability (Cincotta, 2015). Saffron is particularly well suited to Afghanistan's climatic and environmental conditions. Provinces such as Herat, Ghor, Faryab and Badghis possess semi-arid climates, cold winters and well-drained soils that mirror the conditions of neighbouring Iran's Khorasan region, the world's dominant saffron-producing area. Afghan saffron is now internationally praised for its high crocin content and aromatic quality, winning awards from the International Taste and Quality Institute in Brussels (iTQi, 2025). Economically, saffron offers Afghan farmers a high-value, low-volume crop that can be transported and stored far more easily than fresh agricultural produce. While opium poppies have historically provided high profit margins, saffron offers competitive returns without the legal risks or associated violence (Ruttig, 2014). Women in particular have benefited from this: saffron harvesting and stigma separation are labour-intensive tasks frequently done within households, providing income to female workers who often lack access to land or markets (Abirafeh, 2009). The shift from opium to saffron is not without challenges. Issues such as limited processing infrastructure, supply chain inefficiencies and market access barriers still impede the industry's growth (Alizadah *et al.*, 2026). Political instability poses additional threats to long-term investment. Nevertheless, saffron remains one of Afghanistan's most successful agricultural development initiatives. It represents not only an economic alternative but also a symbolic move toward legal livelihoods, community resilience and international integration (Byrd, 2012).

1.3.2 EU saffron markets (Spain, Italy, Greece)

The European Union (EU) is a major producer and consumer of high-quality saffron, with Spain, Italy and Greece being the main producing countries. Although European production is negligible compared to Iran's total production, saffron grown in the EU fetches top prices due to strict quality controls, protected geographical indications and centuries-old cultivation traditions (Freedman, 2008). Spain is the most significant European producer, in particular the region of Castile-La Mancha, whose 'Azafrán de La Mancha' enjoys the prestigious PDO-Protected Designation of Origin status (Dalby, 2000). Spanish saffron is recognized by its brilliant colour and smoked aroma, related to agriculture practices introduced by Moors (Muslim Andalusian agronomists) during the Andalusian period (Willard, 2001). Spain is also among the most important re-exporters in the world, buying raw saffron from Iran and Afghanistan, repackaging it, and selling it under Spanish brands with enormous profit margins (Dalby, 2000). Italy's production is modest but highly valued. Regions like Abruzzo, Sardinia and Tuscany produce their saffron under PDOs, like 'Zafferano dell'Aquila', acclaimed for an intensive chromatic strength (Montanari, 2012). Italian saffron has its roots in the Middle Ages and is linked to the monastic cultivation and local guilds that played an essential part in regional food, such as risotto alla milanese. Small-scale, artisanal production ensures high quality but limits export competitiveness (Montanari, 2012). Greece is proud of the famous 'Krokos Kozanis', a deep red-coloured and PDO saffron with floral overtones. It grows primarily in western Macedonia and has obtained an essential and vital contribution as well as economic importance from local cooperatives (Krokos Kozanis Producers' Cooperative, 2021). The Kozani Saffron Producers' Cooperative is considered in many respects to be a model of successful community-based agricultural management, combining traditional methods with modern quality assurance practices (Ordoudi and Tsimidou, 2004; Kozani Saffron Producers' Cooperative, 2026). Throughout the EU, saffron production is characterized by high labour costs, artisan techniques and rigorous standards for quality – conditions that render premium-market products but, together with Iran, very low production volumes.

1.3.3 Emerging markets: USA and Australia

Interest in specialized plant species, high-value crops and climate adaptability has driven the expansion of saffron cultivation into non-traditional regions in recent years, particularly the USA and Australia. There was significant growth in saffron cultivation in the USA, particularly in the north-east, where Vermont, Pennsylvania and New York had included the spice plant as part of their niche agricultural product mix (Gheshm and Brown, 2021). Research at the University of Vermont has shown that saffron can be grown commercially in temperate climates with good soil management and protection during winter (Ordoudi and Tsimidou, 2004). American saffron growers tend to operate at a small scale and focus on high-margin, direct sales to chefs, local markets and value-added artisanal producers, capitalizing on freshness, organic practices and traceability (Hodgdon, 2020). Australia has become another crucial new saffron producer, mainly in the regions of Tasmania and Victoria, where the cool winters and dry summers produce Mediterranean like growing conditions for saffron (Elliott, 2017). Tasmania's saffron industry is dominated by family-owned farms that have received international awards for their superior-quality stigmas, characterized by high values of the carotenoid pigment crocin (Hosseinzadeh and Nassiri-Asl, 2013a). Australian producers benefit from a strong position in modern agricultural technologies, exemplary food safety conditions and increasing demand for their high-quality spices in regional markets. Both the USA and Australia are small-scale producers of high-quality saffron, and their product reaches the speciality and luxury markets rather than finding its way into mass global supply chains (Gregory *et al.*, 2007).

1.4 Modern Scientific and Medical Research

Saffron has been receiving intense attention in scientific studies over the past decades, especially for its distinct phytochemical composition. Recent studies reveal that saffron contains high amounts of bioactive compounds like crocin, crocetin, picrocrocin and safranal, which are associated with antioxidant, anti-inflammatory, neuroprotective and anti-depressant effects (Hosseinzadeh and Nassiri-Asl, 2013b).

Some of the most interesting findings relate to saffron's use in mood disorders. A number of randomized controlled trials have shown that saffron extracts are as effective as conventional medications like fluoxetine and imipramine in treating mild to moderate depression, without adverse effects (Akhondzadeh *et al.*, 2004; Kazemi *et al.*, 2021). Saffron has been found to affect serotonin reuptake, limit neuroinflammation and increase the antioxidant potential of neuronal tissues in animal studies (Shafiee *et al.*, 2016). Furthermore, it has been found effective in neuroprotection, especially in relation to age-related cognitive impairment and Alzheimer's diseases. Saffron supplementation has been found to provide similar benefits as conventional Alzheimer's medications like donepezil in early stages of the disease (Akhondzadeh *et al.*, 2010a). Experimental studies using saffron compounds like crocin and crocetin have indicated that it can suppress beta-amyloid protein aggregation and protect neuronal cells from oxidative stress (Koulakiotis *et al.*, 2020). The antioxidant effect of saffron is considered one of its best-documented properties. Saffron components such as crocin and safranal help counteract the formation of free radicals, suppress inflammatory processes and protect against oxidative degradation of cellular components (Samarghandian *et al.*, 2017). Due to the diverse effects of saffron, which is used to treat numerous diseases such as cardiovascular, metabolic and neurogenerative diseases, in which oxidative stress plays a part, this property has attracted considerable attention. Preliminary and clinical studies have shown that taking saffron supplementation can have positive effects on the cardiovascular system, for example by lowering blood pressure, serum cholesterol and triglycerides (Kamalipour and Akhondzadeh, 2011). Anti-inflammatory andioxidant property has been suggested as a mechanism for improvement of the endothelium. Further studies also suggested that taking saffron supplementation may help improve insulin sensitivity and glycaemic control; hence, a lot of interest has been directed towards saffron for management of metabolic conditions like diabetes (Asgarpanah and Kazemivash, 2013). Studies in which saffron compounds, particularly crocin, were examined under laboratory conditions suggest that some of them possess anti-cancer properties. Saffron's crocin has been shown to induce cancer cell death, impede tumour development and suppress metastatic spread in animal models (Samarghandian *et al.*, 2017). Furthermore, saffron has also been found effective in treating a range of reproductive disorders such as erectile dysfunction, libido and satisfaction disorders, particularly in cases caused by medication (Modabbernia *et al.*, 2012). In women, it has also been used to alleviate symptoms such as premenstrual syndrome, dysmenorrhoea and mood-associated disorders (Agha-Hosseini *et al.*, 2008). Due to its antioxidant properties, saffron is used in various cosmetic formulations for conditions to treat pigmentation disorders as a brightener, and an anti-ageing agent. From early dermatological literature, there appears to be a potential inhibitory effect of saffron extract on melanogenesis, anti-inflammatory properties, as well as support for wound healing (Rigi *et al.*, 2021). Although the potential benefits of saffron are being unravelled, various issues still hamper commercial production. These problems relate to the expensive production, lack of standard composition due to different origins, and the urgent need for comprehensive clinical studies. The increasing knowledge in this area anchors saffron in a mysterious interface between conventional medicine in the past, present and possibly also in the future.

The current value of saffron demonstrates how a traditional commodity such as saffron, with all its connotations of purity, spirituality and luxury, is evolving to meet contemporary meanings and uses while retaining all its traditional associations with luxury and spirituality (Toussaint-Samat, 2009). The culinary use of saffron has seen a revival with enhanced contemporary interest in traditional foods and cuisine. Luxury restaurants and Michelin-starred chefs continue to use saffron with great interest in their culinary innovations of haute cuisine, fusion cuisine, desserts and other luxury food products (Wright, 1999). This contemporary revival of saffron use in cuisine, in addition to positioning saffron as a traditional use object, further ascertains it as a luxury goods object available globally, with significations of luxury, craftsmanship and exclusivity (Freedman, 2007, 2008).

Saffron holds a great significance for diaspora groups, including Persian, Kashmiri, Mediterranean and South Asian communities. Within Iranian diaspora groups, saffron plays a significant role in celebratory cuisine like sholezard, saffron rice and other sweet dishes that are prepared as part of Nowruz celebrations (Batmanglij, 2011). This trend has been seen among other communities such as South Asians and Middle Eastern groups, where saffron plays a substantial role in wedding ceremonies, religious services and communal meals (Achaya, 1994).

With growing recommendations from scientific research on saffron's advantages as an antioxidant, mood supplement and anti-inflammatory agent, saffron has also found a niche as a natural product especially in high-quality skincare. High-end companies are using saffron extract formulations in anti-ageing and brightening creams, while traditional and contemporary wellness companies promote saffron tea, dietary supplements and nutraceuticals as natural remedies for stress relief and mood regulation (Rigi *et al.*, 2021)

In a world where consumers increasingly demand responsibly manufactured, handcrafted products, saffron has come to represent authenticity. Luxury producers and retailers emphasize their saffron's provenance, from La Mancha to Khorasan and Kashmir (European Commission, 2020). The increased authenticity of saffron has in turn contributed to its establishment in luxury markets worldwide, where rarity and authenticity are key luxury drivers (Bowen, 2015).

The semiotics of saffron still play a role in contemporary literature, film and visual arts today. Here, saffron symbolically represents memory, eroticism, diaspora and pride. On a different note, visual artists are using the colour and significance of saffron in installations that portray identity and diaspora (Naficy, 2001). In this way, saffron becomes more than a culinary object in ritual practices; it becomes a part of cultural discourse in contemporary visual arts.

The contemporary cultural significance of saffron derives from its function as a link between ancient traditions and modern values. Whether in high cuisine, alternative healing, diasporic practices or art, saffron retains meanings from the past even as it translates to a fresh cultural setting. This role of saffron in both everyday life and high cultural spheres indicates that traditional goods can continue to evolve in a globalized world (Freedman, 2008).

1.5 Sustainability and Challenges

Despite its long and storied history, not to mention its current global value, there are a number of environmental, economic and social challenges facing saffron production that continue to affect the sustainability of the industry. For this reason, the future of saffron, whose cultivation is highly labour-intensive, sensitive to climate conditions and based on traditional knowledge, depends on ecological farming methods and fair economic structures. This differs by region but reflects broader global pressures affecting traditional systems of agriculture.

1.5.1 Climate change and environmental stress

Climate change represents one of the most significant threats to saffron cultivation in the world. Increased temperatures, rainfall in the cultivation area and increased frequency of droughts because of climate fluctuations especially threaten major saffron-producing regions like Iran, India (Kashmir), Afghanistan and Spain. Saffron crocus requires cold winters for dormancy and dry summers for flowering; deviations from these conditions can sharply reduce yield and quality (Karimi and Mohammadi, 2018). Water scarcity, especially in Iran and Afghanistan, is particularly critical, with groundwater depleting due to long drought conditions threatening the viability of traditional farming communities (Ghorbani and Koocheki, 2017).

1.5.2 Labour intensity and shrinking workforce

Saffron harvesting remains an exceptionally labour-intensive process, requiring tens of thousands of flowers to produce a single kilogram of dried stigmas. The stigmas must be hand-plucked within hours of flowering, often before sunrise, which demands skilled, coordinated labour. In many places, rural depopulation and the declining interest among younger generations pose long-term challenges to the sustainability of saffron cultivation (Gresta *et al.*, 2008b). Without sufficient manpower, both productivity and quality will be negatively affected.

1.5.3 Adulteration of market and fraud

As saffron is the most expensive spice in the world, it is also one of the most frequently adulterated. Fraud ranges from mixing genuine stigmas with dyed corn (maize) silk or safflower threads to adding artificial colouring or selling diluted powders. These practices undermine producers' income, shake consumer confidence and distort global pricing. Despite modern authentication technologies, adulteration remains a global persistent issue (Ordoudi and Tsimidou, 2016).

1.5.4 Economic volatility and market dependency

Most of the saffron producers have to deal with unstable market conditions because of fluctuating prices, export barriers and intermediaries. International sanctions against Iran further limit the available financial transactions and complicate the export logistics of the world's largest producer. Afghan producers struggle with inconsistent infrastructure, very few processing facilities and extremely fragmented supply chains, all of which inhibit access to the international market (Alizadah *et al.*, 2026). European producers enjoy the protection of PDO regulations but are limited in competitiveness by high labour costs and small-scale production.

1.5.5 Knowledge transmission and genetic resources

Traditional knowledge is at the heart of saffron cultivation, including planting cycles and soil conditioning right through to harvesting and drying techniques. Modernization, urban migration and generational change threaten the transmission of this knowledge (Koocheki, 2004). The genetic base of saffron is limited, as it is a sterile plant propagated by corms, making it susceptible to diseases, climate stress and loss of biodiversity. There is an urgent need for preservation of local landraces and investments in research on the genetic resilience of the plant for long-term sustainability.

1.5.6 Opportunities for sustainable development

Despite such challenges, there is considerable opportunity for saffron to contribute to sustainable development when managed appropriately. With a low water consumption compared to many commercial crops, saffron presents a water-saving alternative to intensive agriculture, particularly in dry areas (Gresta *et al.*, 2008b). Saffron's high value brings pronounced economic opportunities for small-scale farmers, particularly women, who make important contributions to stigma separation and local processing. Certification systems, such as PDO status and organic labels, would protect both producers and consumers by promoting transparency, authenticity and fair-trade practices.

1.6 Wild Saffron Species

The genus *Crocus* comprises approximately 80 species, which are distributed throughout Europe, the Mediterranean region, Central Asia and western Pakistan, with Greece and Turkey recognized as its primary centres of diversity.

Crocus is a perennial plant characterized by a corm that is covered with fibrous, membranous or leathery tunics. The plant lacks a true stem. Its leaves are green, basal and either flat or navicular. They usually develop after flowering, though occasionally they may appear simultaneously with the flowers. The upper leaf surface bears a distinct white stripe, while the lower surface contains several prominent veins running along the sides of a broad or narrow navicular band. The flowers are either vernal or autumnal and may be solitary or clustered. The peduncle is very short and subterranean, sometimes enclosed by a membranous sheathing leaf. The bract is membranous, with the bracteole either of equal size, smaller or occasionally reduced. Flowers are actinomorphic. The floral tube is tall and slender, with six perianth segments arranged in two whorls, which may be equal or nearly equal in size. Stamens are inserted at the throat of the tube and are shorter than the perianth. The filaments are thin or slightly broadened, either hairy or glabrous. The style terminates in three or more stigmas. The ovary is subterranean. The fruit is a capsule, cylindrical to elliptical in shape, usually situated at ground level on a short and narrow base. In autumn-flowering species, the capsules develop and open in the following spring. Seeds are numerous, spherical to elliptical, and brown or reddish in colour (Izadpanah *et al.*, 2014).

Based on morphology, Mathew (1982) divided the genus *Crocus* into two subgenera, subgenus *crocus* with extrorse anthers, and subgenus *crociris* with introrse anthers. The latter subgenus is monotypic, comprising only *C. banaticus*. Subgenus *crocus* was further subdivided into two sections (section *Crocus* and section *Nudiscapus*) including six and nine series, respectively. These sections were defined by the presence or absence of a prophyll, and some essential characters for the delimitation of the series include corm tunic structure, leaf structure, presence/absence of a bracteole, division of the style, anther colour and flowering time (Petersen *et al.*, 2008). The members of the genus *Crocus* have adapted to very diverse ecological conditions, ranging from shady and humid forests to high alpine meadows, rocky outcrops and even harsh desert habitats, and exhibit a wide distribution, spanning from north-western Africa and Portugal along the Mediterranean coasts, eastwards to north-west China. At the same time, northwards covers approximately half of Europe. The Balkan Peninsula and Asia Minor have emerged as the main centres of

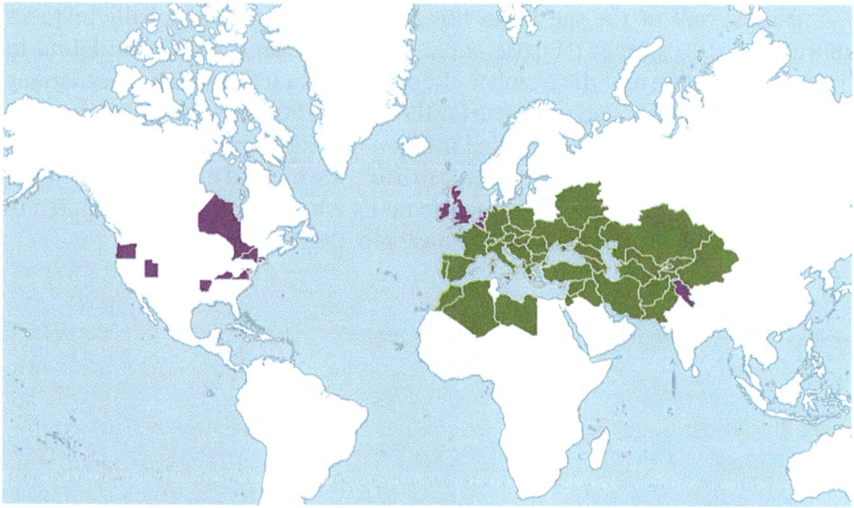

Fig. 1.2. Geographical distribution of wild and cultivated saffron species in the world: native (green); introduced (purple). https://powo.science.kew.org/taxon/urn:lsid:ipni.org:names:20293-1. Figure used under licence CC-BY-3.0.

biodiversity, hosting more than half of the recognized species (Rukšāns, 2017). The taxonomic classification of the genus *Crocus* has undergone substantial revisions over the past half-century (Dolatyari *et al.*, 2024). Back in 1982, when Brian Mathew published his monograph on the genus, he included 80 species and 37 taxa at the subspecies level, which were divided into 15 series based on their morphological characteristics. Intensive research projects on the genus *Crocus* began at the turn of the century, first in Turkey (Türkiye), then in the Balkans, the Pyrenees and Iran. Subspecies are no longer recognized by leading crocus researchers, and all are now considered full species. In 2017, the total number of recognized species reached 235 and 6 years later, in 2023, this number rose to 262 (Rukšāns, 2017, 2023). Discoveries continue to expand our knowledge of *Crocus* diversity (Advay, 2024). According to Rukšāns (2017), 250 accepted species have been listed from the world on Plants of the World Online (POWO) that are native to: Afghanistan, Albania, Algeria, Austria, Baleares, Bulgaria, Central European Russia, Corsica, Crete, Crimea, Cyprus, Czechia and Slovakia, East Aegean Islands, France, Germany, Greece, Hungary, Iran, Iraq, Italy, Kazakhstan, Kyrgyzstan, Lebanon and Syria, Libya, Morocco, North Caucasus, north-western Balkan Peninsula, Pakistan, Palestine, Poland, Portugal, Romania, Sardinia, Sicily, South European Russia, Spain, Switzerland, Tajikistan, Transcaucasus, Turkmenistan, Turkey/Türkiye, Ukraine, Uzbekistan and Xinjiang, and introduced to: Arkansas, Belgium, Connecticut, Delaware, Great Britain, Ireland, Kentucky, Maryland, Massachusetts, Netherlands, New York, Ontario, Oregon, Utah, Virginia and Western Himalayas (Fig. 1.2).

Almost half of the species in the genus *Crocus*, which globally includes more than 200 species (Harpke *et al.*, 2014; Rukšāns, 2017), are found in Turkey. This value and the number of endemic taxa (*c.*120) show that Turkey is an important distribution centre for the genus. Presently, 24 species are recognized in Iran. They are classified into two sections: *Crocus* and *Nudiscapus*. Notably, 18 species (75%) are endemic to Iran, of which 14 are steno-endemics; the remaining four are subendemic elements whose distribution is limited to Iran and the adjacent countries (Yüzbaşioğlu, 2019).

2

Botany: Taxonomy, Morphology and Conservation of Natural Habitats

Hassan Esmaeili and Edris Mahdavi-Fikjvar

2.1 Introduction

Saffron, obtained from the desiccated stigmas of *Crocus sativus* L., is recognized as the most expensive spice globally by weight, selling for up to US$ 3500/kg in international markets (Karbasi and Zandi Dareh Gharibi, 2022). In 2023, the saffron market was estimated to be around US$ 602.2 million worldwide. Strong growth is predicted through 2030 with an average annual growth rate of 7.1% and the potential of exceeding US$ 1 billion by the end of the decade (Acharya *et al.*, 2025). This extraordinary value is due to the intensive manual labour required to harvest the delicate red scars from each flower, with around 150,000 to 200,000 flowers needed (equivalent to around 440,000 hand-picked stigmas) and approximately 400 h of skilled labour needed to yield just 1 kg of dried saffron (Pandita, 2021; Fujii *et al.*, 2022). The global production of saffron exceeds 418 t annually, with Iran contributing 90% of production, followed by India, Spain, Greece, Afghanistan and Morocco (Cardone *et al.*, 2020; Kazemi-Shahandashti *et al.*, 2022). Iran (US$ 102 million) and Spain (US$ 49.8 million) were the top exporters in 2019, when the global trade value exceeded US$ 229 million (Fujii *et al.*, 2022). Beyond agriculture, saffron has a significant economic impact in other industries like cosmetics, gourmet foods (like Italian risotto and Spanish paella), liqueurs, natural dyes and nutraceuticals (Winterhalter and Straubinger, 2000; Mzabri *et al.*, 2019; Cardone *et al.*, 2020). Demand in pharmaceuticals is driven by modern pharmacology's confirmation of its antidepressant, antioxidant, anti-inflammatory and anti-cancer properties (Srivastava *et al.*, 2010; Marrone *et al.*, 2024). In addition, this spice is important to rural livelihoods. For example, its cultivation is a vital source of income in areas like Taliouine, Morocco, where it employs an abundance of female labourers during the harvest (Mzabri *et al.*, 2019).

2.2 Origin and Evolution

Saffron has been grown for more than 3500 years. Evidence of its use traces back to the Minoan civilization, as shown in ancient paintings from the Bronze Age palace of Knossos in Crete (Goliaris, 1999; Negbi, 1999). Saffron had been employed for medicine, cosmetics, textiles and religious ceremonies by the ancient Persians, Egyptians and Romans. During the Middle Ages, its importance extended along trade routes to South Asia, the Middle East and finally to Europe. People appreciated the spice highly not only for its potent colour and taste but also for its medicinal properties, such as sedative, aphrodisiac and as a treatment for stomach problems (Winterhalter and Straubinger, 2000; Fernández, 2004). Saffron persists as a crop that is crucial for both the economy and medicine today. Researchers have confirmed many of its traditional uses by finding active ingredients such as crocin, safranal and picrocrocin, which exhibit antioxidant, antidepressant and anti-cancer properties (Bathaie and Mousavi, 2010; Pitsikas, 2016). Saffron remains a labour-intensive and valuable crop; consequently, understanding its evolutionary history is important for ongoing genetic, agronomic and pharmacological studies aiming to improve its productivity and therapeutic potential.

There are several theories about the origins of the word ‘saffron’. It originates from a rich linguistic and cultural history spanning millennia, with scholarly consensus tracing its roots to ancient Persian. The old Persian word ‘zarparān’, which means ‘gold strung’ or ‘golden leaves/petals’, is the most widely accepted origin (Asgari *et al.*, 2022). This references saffron’s vivid golden-yellow stigma threads or its dyeing properties. Classical Persian dictionaries reinforce the idea that ‘zar’ means ‘gold’ and ‘par’ means ‘petal/leaf’. Through cultural exchange, the word changed to ‘za’farān’ in Arabic and then to ‘safranum’ in medieval Latin. Arabic mediation (‘za’farān’) facilitated its spread to Europe, where it became a part of languages and cultures. The Greeks called saffron krokos, which was later made official in Linnaean taxonomy as *Crocus sativus* (Winterhalter and Straubinger, 2000). Archaeological evidence from Castile-La Mancha, Spain, dating back to around 2400 BCE, demonstrates that saffron was first used as a dye for textiles. Later, Babylonian texts from Hammurabi’s reign (1800–1700 BCE) have documented its culinary application (Dar *et al.*, 2017). The Egyptian medical document known as the ‘Ebers Papyrus’ (Dai *et al.*, 2021) shows that cultivation first definitely happened in Minoan Crete around 1550 BCE.

The saffron plant is recognized for its ancient domestication and its enduring value as a spice and medicinal plant. There are some different opinions about where this plant originates from. Some evidence suggests that saffron was first found on the slopes of the Zagros Mountains in Iran during the Median Era (559–609 BCE). The native inhabitants of that area also established the world’s first saffron fields. The exact geographical and evolutionary origin of saffron has been the subject of extensive scientific investigation. Some reports have declared that saffron emerged from the eastern Mediterranean, maybe from Asia Minor and Persia (Winterhalter and Straubinger, 2000). Evidence of saffron's use dates back 50,000 years to pigment-based cave paintings in Iraq, but its cultivation began during the Minoan civilization (1700–1500 BCE). Frescoes at Akrotiri (Santorini), such as 'The Saffron Gatherers' (see Fig. 5.1C), depict stigma harvesting, suggesting early domestication in Bronze Age Greece. By 300 BCE, Theophrastus documented corm propagation in his book *Historia Plantarum* (Kazemi-Shahandashti *et al.*, 2022).

While Nikolai Vavilov postulated that saffron originated in the Middle East, alternative scholarly hypotheses proposed Asia Minor or the south-western Aegean islands as its cradle (Tammaro, 1999; Dai *et al.*, 2021). Today, it is widely accepted that *C. sativus* originated in the eastern Mediterranean region, particularly in the area encompassing present-day Greece and Crete (Mathew, 1982; Kazemi-Shahandashti *et al.*, 2022).

According to classical morphological studies, *Crocus cartwrightianus* is the closest relative of cultivated saffron (*C. sativus*) (Mathew, 1982). Cytogenetic and molecular evidence suggests that *C. sativus* is an autotriploid derived from *Crocus cartwrightianus*, a wild, diploid species endemic to southern Greece, through autopolyploidization (Schmidt *et al.*, 2007; Nemati *et al.*, 2019).

The genetic diversity of global saffron accessions has been extremely low, which indicates that a single clonal lineage has been grown around the world (Busconi *et al.*, 2015). Persistent uncertainties surround saffron's evolutionary mechanisms, reflecting ongoing scholarly disagreement about its genetic origins. There has been a lot of debate regarding the phylogenetic origins of *Crocus sativus*. Some studies have employed molecular techniques to determine the evolutionary background and phylogeny of saffron (*Crocus sativus* L.), a sterile triploid species that is unable to reproduce sexually. Initial molecular analyses led to conflicting hypotheses about its ancestor species, as early markers lacked enough resolution. Frello *et al.* (2004) detected minimal divergence in plastid DNA sequences among *C. sativus*, *C. cartwrightianus*, *C. thomasii* and *C. hadriaticus*, preventing clear ancestral assignment. Further investigations have evaluated chloroplast, ribosomal and nuclear loci (Petersen *et al.*, 2008; Erol *et al.*, 2014) and were unable to find a clear domestication locus or progenitor. RAPD analyses suggested *C. cartwrightianus* and *C. thomasii* as closest relatives, while *C. pallasii* and *C. asumaniae* were more divergent and more distant relatives of saffron (Caiola *et al.*, 2004). While one study supported *C. cartwrightianus* as the sole ancestor (Gismondi *et al.*, 2013), others suggested that *C. hadriaticus* or *C. thomasii* were also involved in the formation of the hybrids (Tsaftaris *et al.*, 2011). Applying advanced genotyping, based on GBS SNP data (Nemati *et al.*, 2018) it was proposed to place *C. sativus* inside the species *C. cartwrightianus* and to exclude other taxa.

Integrated genomics (nuclear loci, chloroplast genomes and flow cytometry) performed by Nemati *et al.* (2019) confirmed wild *C. cartwrightianus* populations from south of Athens, Greece, as the exclusive progenitor through autopolyploidization. While earlier studies proposed an allotriploid hybrid origin (*C. cartwrightianus* × *C. pallasii* subsp. *pallasii*) (Alsayied *et al.*, 2015), genomic evidence now establishes saffron as an autotriploid derived exclusively from *C. cartwrightianus* (Nemati *et al.*, 2019). Genetic analyses, including genotyping-by-sequencing (GBS) and fluorescence *in situ* hybridization (FISH), confirm that 99.3% of saffron alleles derive from *C. cartwrightianus*, between 1600 and 350 BCE (Nemati *et al.*, 2019; Salami and Husaini, 2022). This domestication probably happened when people selected the most promising *C. cartwrightianus* mutants with longer stigmas (Francis and Ramandi, 2020; Asgari *et al.*, 2022). It is fixed through vegetative propagation by corms, since saffron plants are male sterile. Consequently, all global saffron accessions today represent a single clonal lineage with minimal genetic diversity.

The evolutionary transition from wild *Crocus* species to cultivated *C. sativus* involved both natural hybridization and artificial selection. The process of domesticating saffron probably started with *Crocus cartwrightianus*, a wild flowering plant. A mutant form of this plant with long stigmas had been discovered, grown and officially domesticated in Minoan Crete during the Late Bronze Age. Unlike its wild progenitor, *C. sativus* is sterile and cannot reproduce sexually. It propagates exclusively through vegetative division of corms, a feature that has led to genetic uniformity across cultivated populations (Gresta *et al.*, 2008b). This clonal reproduction ensures the conservation of desired traits such as high crocin content but also restricts genetic diversity, making the plant vulnerable to environmental stress and disease. To overcome these limitations, modern efforts in biotechnology and molecular breeding have focused on identifying genetic markers and improving propagation techniques (Ahrazem *et al.*, 2015).

2.3 Distribution

The genus *Crocus* is the most diverse in the Balkans and Turkey. Its species can be encountered in meadows, scrub and woodlands across a wide range, from western Europe and north-western Africa to the Near East and Central Asia to western China (Moraga *et al.*, 2009; Harpke *et al.*, 2015). The Balkan Peninsula and Asia Minor show the greatest biodiversity (Mathew, 1982). Saffron (*C. sativus*) grows mostly in a latitudinal belt (30–50°N) between the Mediterranean and Kashmir, extending from 10°W to 80°E (Yasmin and Nehvi, 2013). Historical documents indicate that the first cultivation occurred near Iran's Zagros and Alvand mountains during the Median Kingdom (708–550 BCE) (Pandita, 2021), though specific uses remain undocumented (Abrishami, 1997; Dadkhah *et al.*, 2003). According to Dadkhah *et al.* (2003), the ancient Persians were the first to cultivate saffron specifically rather than relying solely on wild sources. Saffron stigmas were introduced to China during the Yuan Dynasty (around the 14th century CE). The Chinese Pharmacopoeia outlines them for modulating haemodynamic parameters (blood circulation enhancement) and neuropsychiatric functions (depression attenuation) (Xu *et al.*, 2024). The most definitive iconographic evidence that saffron was grown in earlier times comes from Aegean Bronze Age artworks (1700–1500 BCE) (Deo, 2003; Dewan, 2015), especially those from Minoan culture. Primary archaeological sites documenting these representations are located on Crete and Thera (modern Santorini), both in present-day Greece (Warren, 2005).

Although saffron is cultivated in numerous countries and has also been introduced to several new regions (Fernández, 2004), Iran, with its long history and significant production, remains the world's leading producer (Fernández, 2004; Ghorbani, 2008; Ahmad *et al.*, 2011). Several other nations are notable producers, while cultivation also occurs on a smaller scale in various other locations (Abdullaev, 2002; Hill, 2004; Keify and Beiki, 2012). Most global saffron farms are concentrated in the Mediterranean region and in Iran, Turkey and Jammu and Kashmir, India (Lim, 2012). Iran remains the primary cultivator, with key provinces being Khorasan Razavi, South Khorasan, North Khorasan and others (Gupta *et al.*, 2021). India has the second-largest cultivation area but low productivity (2.0–2.5 kg/ha), mainly in districts of Jammu and Kashmir. Afghanistan's saffron industry has expanded rapidly as an alternative to opium, supported by international programmes (Mollafilabi and Aslami, 2020). Spain's production, centred in Castile-La Mancha, saw a reduced area but increased yield from 1997 to 2016. Other producers include Greece (mainly in Kozani), Italy, Morocco, Turkey, Azerbaijan and China (Fernández, 2004; Ünaldı, 2007).

In trade, Iran, Spain, Afghanistan, Greece and France are leading exporters, while Spain, Hong Kong, the USA, Italy and India are top importers (Gupta *et al.*, 2021). Recently, China, Australia and the USA have initiated small-scale or experimental farming using controlled environments and improved propagation to meet local demand (Ahrazem *et al.*, 2015). Although *C. sativus* adapts to various climates and reproduces via corms, it remains sensitive to soil, irrigation and temperature. Thus, global cultivation expansion depends heavily on local conditions and labour availability for harvesting. Given rising demand in culinary, nutraceutical and pharmaceutical markets, understanding saffron's ecological and economic dimensions is vital for sustainable development and trade (Fernández, 2004; Gresta *et al.*, 2008a).

2.4 Taxonomy

Saffron (*Crocus sativus* L.) is a triploid ($2n = 3x = 24$) and sterile species that belongs to the Iridaceae family, comprising a diverse group of flowering plants known for their ornamental and medicinal properties. Within this family, saffron is placed in the genus *Crocus*, which includes approximately 85–100 species distributed mainly across Europe, North Africa and parts of Asia. It is monocotyledonous, evidenced by its single-cotyledon embryo and parallel leaf veins (Saxena, 2010). Saffron is one of the few intentionally cultivated species in the genus, valued for its culinary properties. Historical official designations for saffron (e.g. *crocus Anglicus*, *crocus Hispanicus*, *crocus Gallicus*) denoted geographical origins and should not be conflated with Linnaean binomial nomenclature, despite superficial similarities in form (Francis and Ramandi, 2020). *Crocus sativus* has taxonomically assigned to the series *Crocus* (the saffron group) based on shared morphological characters, alongside several congeneric species. Among these, *C. cartwrightianus* Herbert exhibits the closest morphological affinity to saffron. *C. cartwrightianus* is a fully fertile taxon that grows in rocky, open areas from the coast to the submontane elevations (1000 m) across Greece (Francis and Ramandi, 2020). Taxonomic delineation relies on cytology (chromosome counts), stigma length and floral anatomy, as molecular markers like RAPD and ISSR confirm minimal interspecific divergence (Saxena, 2010). The taxonomy of *C. sativus* poses unique challenges due to its sterility, clonal propagation and lack of genetic recombination, which limit the natural variability observed in wild taxa. Unlike sexually reproducing species, *C. sativus* is propagated exclusively via corms, resulting in a genetically uniform population worldwide. This raises issues in taxonomic delineation and biodiversity conservation, as conventional morphological and reproductive traits used in plant systematics are absent or limited (Fernández, 2004; Gresta *et al.*, 2008b). Taxonomic identification in saffron relies on various methods, including corm tunic architecture and stigma/style length ratios (Saxena, 2010). Karyotype analyses, such as fluorescent *in situ* hybridization (FISH), have been crucial in unravelling saffron's genetic makeup (Schmidt *et al.*, 2019).

Nevertheless, advances in molecular phylogenetics, including the use of ISSR, RAPD and SSR markers, have helped clarify its taxonomic position. Taxonomically, the genus *Crocus* is divided into sections based on morphological traits such as corm tunic structure, chromosome number and flower morphology. *Crocus sativus* is placed in section *Crocus*, a group that includes many autumn-flowering species (Mathew, 1982). The classification continues to be refined with molecular data, which support the monophyly of major clades within the genus and assist in resolving taxonomic ambiguities in closely related species complexes. Understanding the taxonomy of *C. sativus* is critical not only for botanical research but also for breeding programmes, conservation planning and the authentication of commercial saffron products in global markets.

2.5 Morphological Characteristics

Crocus sativus L. is a perennial herbaceous geophyte plant whose height may reach up to 25 cm. It is characterized by a distinctive morphology shaped by both evolutionary processes and domestication for its valuable stigmas. The plant propagates exclusively vegetatively through an underground storage corm that is round, tunicate and about 10–15 mm in diameter (Darvishi *et al.*, 2006; Gresta *et al.*, 2008b). The corm functions as a perennating structure and stores nutrients necessary for sprouting and flowering in the next growth cycle. It is covered in fibrous tunics, which form a net-like protective layer. In healthy specimens, this layer is usually brownish-white (Saxena, 2010).

Corm activation initiates bud emergence, producing 3–5 cataphylls (5–25 cm) that sheath developing shoots. Each sprout has 5–11 linear nomophylls, characterized by being 1.5 to 2.5 mm wide and 50 cm long. They have a white band on the adaxial side, a keel on the abaxial side and stomata that are only found in lateral grooves (Ghorbani and Koocheki, 2017). Leaves persist from autumn to late spring, using photosynthesis to construct corm reserves before summer dormancy. *C. sativus* lacks a true aerial stem like many other members of the Iridaceae family. Instead, the leaves are crucial for photosynthesis and storing energy for the next season's corm development (Fernández, 2004; Gresta *et al.*, 2008b). The plant reproductive structure is the most economically important feature. Cataphylls conceal the bracts and bracteoles where the flowers develop. Each stem leads to one to three flowers with six violet, veined tepals that are fused into a base tube (Delgado *et al.*, 2006). The androecium has three yellow sagittate anthers (15–20 mm) on filaments that are attached to it. The gynoecium features a filiform style terminating in three commercially harvested stigmas (25–32 mm), which are disproportionately long, trumpet-shaped, scarlet structures that typically rest on tepals post-anthesis (Delgado *et al.*, 2006). The flowers bloom in autumn and last only a few days. The stigmas are the spice saffron, which is picked and dried immediately after flowering. These stigmas stick to the style, making the 'thread' of saffron used commercially. Apocarotenoids like crocin supply the bright red colour, and safranal and picrocrocin provide the smell and taste. The stigmas, though morphologically functional, lack viable ovules, and pollen germination fails on receptive surfaces (Chichiriccò and Caiola, 1982; Alsayied *et al.*, 2015). This genetic uniformity limits phenotypic diversity but ensures consistency in spice quality across global accessions.

Saffron corms produce two structurally and functionally distinct root types at the third basal internode: numerous fine, white fibrous roots (approx. 1 mm thick) emerging in a single ring at the corm base for absorption, and large, whitish, tuber-like contractile roots (2 cm diameter, 9–15 cm long) that facilitate optimal soil depth positioning via pulling/pushing activity (Ghorbani and Koocheki, 2017). A potential third type, thinner contractile-absorbent roots appearing later, has been proposed, but its function is unclear (Negbi, 1990).

Corms are spherical storage organs that weigh between 2 and 50 g and are covered by fibrous, reticulate tunics (Fig. 2.1). In the second year after flowering, each mother corm produces 3–4 daughter corms. In the third year, each daughter corm generates 1–6 daughter corms, for a total of about 20–22. From the fourth year on, as older generations decay, the number of corms remains at 18–20. After the flowers bloom in November, new corms grow and reach their full size before May. Every year, new corms grow on top of old ones, which causes a 1–2 cm upward creep. To keep the plants from getting too crowded and exposed to the surface, they need to be lifted and

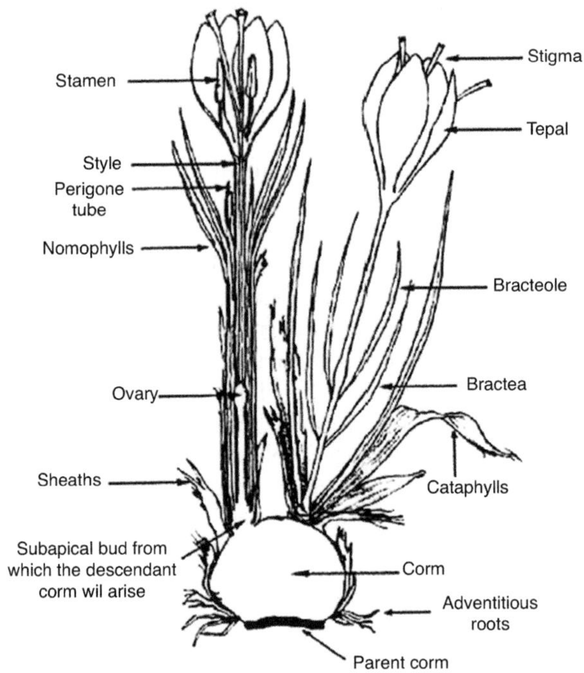

Fig. 2.1. Schematic representation of saffron (*Crocus sativus*) plant organ structure (Alonso *et al.*, 2012). Figure used with permission from Elsevier.

replanted every 4 years (up to 12 years in dryland) (Negbi, 1999; Orti *et al.*, 2004; Kafi, 2006).

Saffron corms do not require a cold period for dormancy termination or floral development (Rees, 1992; Dole, 2003). In autumn (October to November, when the temperature is between 10 and 14°C for Persian saffron), flowering occurs. After that, vegetative growth and the formation of replacement corms happen. Leaf senescence and dormancy onset occur in April–May. High temperatures (the best range is 23–27°C) break dormancy in buds and initiate flowering; however, prolonged exposure delays flowering. Storage at high temperatures can potentially extend the flowering period. Research focuses on controlling flowering to promote hysteranthous leaf appearance (Abarghooie *et al.*, 1999; Molina *et al.*, 2004a, b, 2005; Delgado *et al.*, 2006). As mentioned above, the cormlets are formed around the mother corm during the growing season and serve as the primary means of plant multiplication. This clonal propagation results in genetically identical plants, limiting genetic variation and posing challenges for breeding and disease resistance. Understanding these morphological features is essential not only for saffron cultivation but also for breeding programmes, conservation efforts and post-harvest quality management.

2.6 Ecological and Adaptive Features

Crocus species are mostly found in the Mediterranean region, but they also grow in the Irano-Turanian floristic region to the east. Both areas exhibit a climate with cool to cold winters, and most of the precipitation occurs in autumn, winter and spring, followed by warm, arid summers. The Irano-Turanian Region possesses colder winters and less precipitation overall. Saffron is physiologically adapted to this seasonal regime, wherein the plants proceed dormant from autumn to late spring and survive the summer drought by developing a compact corm (Ghorbani and Koocheki, 2017). *Crocus sativus* L. exhibits a unique set of ecological and adaptive characteristics that allow it to grow in a range of agroecological environments. It grows best in places with cold winters, dry summers and well-drained soils (Gresta *et al.*, 2008b; Koocheki *et al.*, 2016). It can grow in places with climates similar to those of the Mediterranean, like Iran, India (Kashmir), Morocco, Greece and Spain. New efforts have also been initiated to grow it in Afghanistan, Italy and Australia. The optimal growth of saffron (*Crocus sativus* L.) occurs under temperate, dry conditions with specific seasonal patterns, wherein vegetative growth starts during cold temperatures, with optimal conditions comprising rainy autumns, moderate winters and pleasant summers (Fernández, 2004; Mollafilabi, 2004). The species demonstrates significant cold tolerance, surviving temperatures as low as −18°C (Mollafilabi, 2004).

Climate variability, particularly temperature, significantly impacts saffron yield. In Khorasan, Iran, climatic changes over a decade explained 31–66% of yield variation across main cultivation areas, with maximum and minimum monthly temperatures being the primary drivers in most locations (Hosseini *et al.*, 2008). Overall, precipitation had a minor impact and only significantly affected yields in Torbat-e-heidarieh (Hosseini *et al.*, 2008). Researchers found that the temperatures in the spring and early summer had the greatest impact on yield (Hosseini *et al.*, 2008). Farooqi and Sreeramu (2004) declared that adequate precipitation during winter is beneficial for corm cultivation.

The climate of Southern Khorasan (Ghaen, Gonabad, Ferdows and Birjand), characterized by maximum temperatures ≤20°C and minimums reaching 0°C during October–December, represents typical saffron requirements (Kafi, 2006). Flowering is influenced by air temperature and rainfall during spring and autumn, with warm springs and long autumns promoting early flowering. Optimal flowering requires day temperatures of 15–20°C and night temperatures of 6–8°C, with sunny days favouring higher yields. The role of temperature in flower bud differentiation remains unclear. There have been inconsistencies in the research about how temperature affects corm storage and forcing on flowering. For example, Plessner *et al.* (1989) found that the same number of flowers appeared under uncontrolled (~15°C) and controlled (17/12°C day/night) conditions, though storage details were incomplete.

Saffron is a geophyte, which means it can survive adverse conditions by storing energy in an underground organ. The corm enters dormancy during the hot summer months and resumes growth in autumn, initiating the emergence of floral and vegetative organs. This seasonal dormancy is an effective survival mechanism that helps the plant conserve resources and avoid damage from heat or drought (Fernández, 2004). The plant's phenological plasticity, such as flowering in autumn and leaf senescence in the spring and summer, meets well with the changing seasons in Mediterranean and steppe environments. Saffron is relatively drought-tolerant, requiring minimal irrigation during its active growth phase (typically between October and April in the northern hemisphere). However, the quantity and timing of water availability significantly affect corm development and floral yield. Optimal saffron production depends on moderate rainfall or controlled irrigation during corm sprouting and floral induction, while excess moisture or waterlogging can lead to fungal diseases and corm rot (Gresta *et al.*, 2008a; Koocheki *et al.*, 2016). Its low nutrient demand, combined with a tolerance to marginal soils, makes it suitable for cultivation in resource-poor regions, particularly when sustainable practices such as organic mulching and crop rotation are applied.

The quality of saffron is predominantly determined by the levels of the metabolites crocin, picrocrocin and safranal. Safranal, constituting approximately 70% of saffron's volatile fraction, is the principal contributor to its aroma and is formed through the decomposition of precursor glycosides (Kyriakoudi *et al.*, 2015a). Geographical factors such as climate, edaphic properties, topography and biotic interactions have an immense impact on the levels of stigma glycosides, especially safranal (Rezvani-Moghaddam, 2020). Variations in key volatile compounds (safranal, isophorone, 4-keto isophorone, c-pyronene, dihydrooxophorone) are primarily attributed to cultivation site and post-harvest stigma dehydration protocols (Macchia *et al.*, 2013). The altitude has a significant influence on the crocin content (Lage and Cantrell, 2009; Lage *et al.*, 2010), wherein Alpine regions in Italy yield superior quality saffron in terms of crocin, picrocrocin and safranal content (Giorgi *et al.*, 2017; Caser *et al.*, 2018). Giorgi *et al.* (2017) attribute this quality to harvesting closed flowers during early morning hours (minimizing light degradation of crocins), optimized drying and storage. The apocarotenoid content (governed by ISO quality rules) is also strongly influenced by sowing time, confirming the critical role of environmental conditions in accumulating these secondary metabolites responsible for quality (Gresta *et al.*, 2008a).

Agricultural interventions also affect quality. For example, arbuscular mycorrhizal symbionts increase the levels of bioactive compounds (picrocrocin, crocin II and quercitrin) and antioxidants (Caser *et al.*, 2018, 2019). Soil factors, such as the potassium content and electrical conductivity, exhibit a negative correlation with safranal content (Rezvani-Moghaddam, 2020). Controlled environment production consistently yields higher quality saffron (in crocin, picrocrocin and safranal) compared to field cultivation, which could be attributed to optimized climate and nutritional control, potentially utilizing hydroponic substrates and foliar feeding (Poggi *et al.*, 2009; Mollafilabi, 2014).

Saffron exemplifies biochemical and phenological adaptation to Mediterranean climates, balancing water conservation, thermal resilience and biotic defence through specialized ecotypes. Its hysteranthous growth, corm-based resource storage and apocarotenoid chemistry comprise an integrated adaptive framework. However, genetic uniformity heightens vulnerability to pathogens and climate shifts. In summary, saffron's ecological and adaptive traits, such as its seasonal dormancy, drought tolerance, low input requirement and efficient resource use, make it an ideal crop for marginal lands and environmentally sensitive regions. Understanding these features is essential for optimizing saffron cultivation under the pressures of climate change, soil degradation and water scarcity, and for expanding its production to new geographical zones.

2.7 Iranian Wild *Crocus* Distribution

In Iran, eight species of this genus have been identified.

2.7.1 *Crocus almehensis* Brickell & Mathew

In this subspecies, the corm is leathery, and the leaves appear simultaneously with the flowers in spring, 3 to 5 times as long as the flowers, 2–5 mm wide and greyish-green in colour. Flowers 1–3 in number, light yellowish orange in colour (Fig. 2.2). Style the same size as the stamens, with three stigmatic branches at the end, reddish-orange. Capsule 1.7–2.1 cm long, elliptical oblong. Seeds are 2–3 mm in diameter and dark brown in colour. Flowering time: late winter to early spring.

2.7.2 *Crocus biflorus* Mill

In this species, the corm has a leathery to more or less papery coating. Leaves at the same time as flowers in spring, 3–8 in number, 1–3 mm wide, green to greyish-green. Flowers 1–4 in number, white to blue-green. Style often yellow, with three yellow to reddish-orange or rarely white stigmas, equal to or longer than stamens (Fig. 2.3). Capsule cylindrical, 1–1.5 cm long. Seeds are more or less spherical, reddish-brown in colour, with terminal appendage. Flowering time: late March–late May.

2.7.3 *Crocus michelsonii* B. Fedtsch

In this species, the corm is covered with dense reticulate coarse fibres left over from older layers. Leaves appear simultaneously with flowers in spring, 4–7 in number, 1–2 mm wide, more or less equal to the flowers, greyish-green. The flowers are white with a bluish or light purple tint, especially on the outer surface (Fig. 2.4). Style white with a three-branched stigma equal to or slightly longer than the anthers. The capsule is 1.6 cm long, elliptical, and the seeds

Fig. 2.2. *Crocus almehensis* Brickell & Mathew (photograph by Edris Mahdavi).

Fig. 2.3. *Crocus biflorus* Mill. (photograph by Edris Mahdavi).

are about 4 mm in diameter, more or less triangular, shiny brown with a wrinkled shell. Flowering times: February to March.

2.7.4 *Crocus caspius* Fisch. & C.A. Mey

The plant's corm has a light-brown, papery coat with vertical slits. It has 2–9 leaves, 1–2 mm wide and 3.5–9.3 mm long, appearing at the same time as

Fig. 2.4. *Crocus michelsonii* B. Fedtsch (photograph by Edris Mahdavi).

flowering in autumn. Flowers are 1–2, rarely 3. Their colours are white to light blue. The length of their tube is 4–18 cm. Petals are more or less equal, 2–4 cm long and 6–15 mm wide, obovate, more or less acute to obtuse at the end. The colour of stem is orange-yellow, with 3 segment stigmas, considerably taller than stamens; capsule *c.*1–1.3 cm long, more or less orbicular (Fig. 2.5). Seeds *c.*3 mm in diameter, more or less orbicular to slightly angular, dark brown. Flowering time: October–November. This species occurs in the Caucasus and in northern Iran in the Gilan and Mazandran provinces.

2.7.5 *Crocus sativus* L

In this species, the corm is covered with coarse and reticulate fibres. Leaves at the same time as flowers in autumn, 5–11 in number, 1–2 mm wide, greyish-green in colour. Flowers 1–4 in number, purple with dark veins towards base. Stems reddish-orange, with three dark red stigmatic branches longer than the anthers. Flowering time: October–November.

2.7.6 *Crocus speciosus* M.B. subsp. *speciosus*

In this species, the corm is papery and leathery. Leaves asynchronous with flowers but synchronous with fruit in spring, 3–4 in number, 3.5–5 mm wide, green in colour. Flowers are autumnal, often single, bluish-purple, with a light-yellow base and blue purple veins. Stems yellow or orange, with numerous orange stigmatic branches longer than or equal to the anthers (Fig. 2.6). Capsule is 2–2.5 cm long, elliptic. Seeds are 3 mm in diameter, spherical, reddish-brown. Flowering times: September–October.

Fig. 2.5. *Crocus caspius* Fisch. & C.A. Mey. (photograph by Edris Mahdavi).

Fig. 2.6. *Crocus speciosus* M.B. subsp. *speciosus* (photograph by Edris Mahdavi).

2.7.7 *Crocus pallasii* Goldbach subsp. *haussknechtii* (Boiss. & Reut. ex Maw) Mathew

In this subspecies, the corm is covered with dense, reticulate fibres with a long neck of old fibres. Leaves asynchronous with flowers and with fruit in spring, 9–17 in number, 0.5–2 mm wide, greyish-green in colour. Flowers persistent, 2–5 in number, white with light purple to dark blue veins with yellow or white colour at base. Style orange, with three red bright orange stigmas, often shorter or the same length as the stamens (Fig. 2.7). Capsule is elliptical to cylindrical, 2.5–3 cm long. Seeds are 3–4 mm in diameter, spherical and reddish-brown. Flowering time: October–November.

2.7.8 *Crocus cancellatus* Herb. subsp. *damascenus* (Herb.) Mathew

In this subspecies, the corm is covered with coarse, reticulate fibres with a long neck measuring 6 cm. Leaves appear asynchronously with the flowers in spring, 3–5 in number, 2–3.5 mm wide, greyish-green in colour. Flowers persistent, 1–3 in number, white to blue in colour, yellow or white at the base with darker purple veins. Stems yellow or orange with numerous yellow to orange stigmatic branches, shorter than or as long as the anthers (Fig. 2.8). Capsule oblong or cylindrical, 2–3 cm long, seeds 4–6 mm long, reddish-brown in colour. Flowering time: persistent.

2.7.9 *Crocus gilanicus* Mathew

In this species, the corm is covered with thin fibres. The leaves appear asynchronously with flowers in spring. Number of leaves is 4. The flowers are often solitary, white with purple veins. The styles are cream-coloured, with 3–4 stigmatic branches, each equal to or slightly longer than the creamy-white stamens (Fig. 2.9). Flowering time: October.

Fig. 2.7. *Crocus pallasii* Goldbach subsp. *haussknechtii* (Boiss. & Reut. ex Maw) Mathew (photograph by Edris Mahdavi).

Fig. 2.8. *Crocus cancellatus* Herb. subsp. *damascenus* (Herb.) Mathew (photograph by Edris Mahdavi).

Fig. 2.9. *Crocus gilanicus* Mathew (photograph by Edris Mahdavi).

2.8 Conclusion

Saffron, one of the most valuable medicinal and culinary plants known to humanity, holds exceptional historical, economic and scientific significance. Revered for over 3000 years, its value extends beyond its use as a spice, encompassing pharmacological, cosmetic and industrial applications. Saffron is the most expensive spice in the world by weight, which is essential to the economies of countries like Iran, India, Greece and Morocco, and has attracted more attention in recent producers like Afghanistan and China. The bioactive compounds in its stigmas (crocin, picrocrocin and safranal) contribute to its therapeutic effects and market value, driving global demand and research interest (Bathaie and Mousavi, 2010; Pitsikas, 2016). Taxonomically, saffron is a triploid and sterile member of the Iridaceae family, believed to have evolved through natural or induced polyploidy from *Crocus cartwrightianus*, a wild species native to Greece. Its inability to reproduce sexually and reliance on vegetative propagation through corms result in genetic uniformity and pose challenges for breeding and conservation. Despite these limitations, advancements in molecular phylogenetics, including DNA barcoding and marker-assisted studies, have helped clarify its classification and evolutionary origin, reinforcing its close affinity to wild progenitors (Ahrazem *et al.*, 2015).

Morphologically, saffron is a geophyte with specialized features, such as a perennial corm and seasonal phenology, that make it highly adapted to semi-arid, Mediterranean-like climates. Its distinct floral structures, especially the red stigmas, define both its economic value and reproductive limitations. The plant exhibits remarkable ecological and physiological adaptations, including drought tolerance, seasonal dormancy and low nutrient demands, making it suitable for marginal lands and low-input farming systems. These traits also position saffron as a promising crop for sustainable agriculture under climate change scenarios (Gresta *et al.*, 2008b; Koocheki *et al.*, 2016). The global distribution and adaptive success of *C. sativus* reflect not only its cultural and economic importance but also its potential as a model for studies in domestication, clonal propagation and resilience. Ongoing research into its genomics, stress physiology and agronomy continues to offer opportunities for improving yield, quality and environmental resilience. Thus, saffron stands at the intersection of traditional knowledge and modern science, as an ancient plant whose future relevance is shaped by innovation, sustainability and global appreciation.

3 Breeding and Cultivation: Traditional and Modern Cultivation Methods, Harvest and Post-harvest Processing

Hassan Esmaeili

3.1 Introduction

Saffron (*Crocus sativus* L.) is considered the most expensive spice worldwide. This has been mainly due to its secondary metabolites such as crocins, picrocrocin and safranal, for which various health benefits have been identified, including antioxidant, anti-cancer and neuroprotective properties, among others. These promising reports have led to a global increase in demand for saffron worldwide, creating a need for its cultivation and improvement using evidence-based scientific techniques. Until 2000, Iran, Spain and Greece were the most important exporting countries. However, Afghanistan, China, the Netherlands and Portugal were the most important export countries. Recently, however, Afghanistan, China, the Netherlands and Portugal have also gained influence in the market. In 2018, Iran was the largest producer, with a harvest of 404 t from a total area of 111,000 ha, mainly in the province of Khorasan.

In 2014, world production of saffron stood at 301 t, with Iran producing the lion's share of 280.6 t, over 80% of the world's total output. In general, saffron is considered a high-value, low-yield crop, which is why its cultivation is economically indispensable for farmers in arid regions. Improved growing techniques that increase productivity and stigma quality become matters of sustaining rural livelihoods and meeting market demands. Growth is hampered due to the triploid and sterile nature of saffron, which restricts viable seed production and genetic diversity, while its propagation is hugely dependent on corm multiplication, which is a slow process and also susceptible to various soil-borne diseases. These conditions in growing regions such as Iran, India and Spain are further worsened by irregular rainfall, soil degradation and climate change. These problems are further exacerbated by labour-intensive harvesting and stigma separation methods, as these tasks cannot be fully mechanized. For breeding purposes, the lack of sexual reproduction impairs the efficiency of traditional hybridization. Nevertheless, new molecular techniques such as marker-assisted selection, transcriptomics and genome editing offer prospects for overcoming these limitations, even though their practical implementation is still in its infancy and detailed genomic information is still largely lacking.

3.2 Traditional and Modern Cultivation Methods

For the propagation of saffron, which is still done manually worldwide, high-quality corms with a diameter of more than 2.5 cm and a weight of 5–10 g are generally used, whereby the corms must be free of viruses and other pests (Khorramdel *et al.*, 2015). According to Douglas *et al.* (2014) flower and stigma production peaks every 3 years, while the Iranian Ministry of Agriculture recommends a 4-year crop cycle. However, within 4–5 years, yields decline due to competition for nutrients and water, infestation by saffron latent virus (SaLV), which was first reported in Iran by Parizad *et al.* (2017), and the displacement of saffron corms to the soil surface.

Saffron flowers and corms are quite delicate, which makes them unsuitable for mechanical cultivation. This means that saffron farming practices are mostly done manually. Weeding in a saffron plantation is very difficult, as saffron does not grow tall and cannot crowd out other plants. Moreover, weed pressure increases when saffron flowers in autumn and sprouts leaves in spring. Only when the leaves fall off in spring can all the weeds be removed (Parizad *et al.*, 2017). SaLV poses a new threat in that it is not currently causing any damage, but has a potential negative impact on yields and quality in the near future (Parizad *et al.*, 2017). These conditions that make it difficult to cultivate saffron using mechanical farming methods make it ideal for production in countries with low labour costs such as Iran, India and Afghanistan. To a certain extent, the introduction of appropriate cultivation methods such as higher plant population, appropriate crop rotation, the use of virus-free corms and perennial plants with a shorter life cycle of less than 5 years would lead to a more efficient use of resources.

The value of the global saffron market stood at around US$ 602.2 million in 2023. The market was projected to expand with a compound annual growth rate (CAGR) of 7.1% from 2024 to 2030, reaching a value of US$ 1 billion in 2030. This expensive spice provides immense opportunities for small and medium scale farmers globally (Acharya *et al.*, 2025). In the US, the demand for saffron has also risen steadily in recent decades, as it is used as a culinary spice and has proven medicinal properties (Siddique *et al.*, 2020). This can be seen in the volume of saffron imports to the US, which have risen steadily from US$ 5 million (14,000 kg) in 2002 to over US$ 20 million (103,000 kg) in 2022 (Acharya *et al.*, 2025).

Iran is known to be the major saffron producer, with over 80% of total saffron production, but continuous political instabilities, climatic conditions and supply chain issues pose threats to worldwide saffron availability. This provides a justification for identifying a new saffron production site for future availability.

As already mentioned above, Iran remains the largest producer of saffron, but yields have been steadily declining since 2000, despite the expansion of cultivation areas to 21 provinces, highlighting the urgent need for innovative approaches (Khajeh-Hosseini and Fallahpour, 2020).

Some of the most pressing issues are also lack of access to quality corms, input management, lack of genetic improvement and conservation, adoption of traditional production techniques, loss of knowledge, lack of mechanization, growing in unsuitable cold/wet conditions, drought, irrigation and lack of profitability, with 75% of this crop being from small, unprofitable farms, and 80% of these being below 1 ha in size for most of these growers (Koocheki *et al.*, 2017).

Average yields per hectare are also well below their potential range of 10–15 kg, which negatively affects family incomes. In order to improve the situation, mechanized modernization of production techniques was advocated. Upgrading techniques for cultivation, processing and production of high-quality saffron is also needed for the producers.

There are also new approaches that would contribute to increasing yields such as cultivation under controlled conditions using plastic tunnels, greenhouses and forcing chambers, as well as soil-free cultivation, such as in hydroponic and aeroponic systems, among others.

Controlled environment cultivation offers another alternative to traditional outdoor farming practices for counteracting the limitations imposed by unsuitable climatic conditions (Koocheki and Seyyedi, 2016; Khajeh-Hosseini and Fallahpour, 2020). This has become particularly important as saffron has a very high value, but with increasing labour costs, it becomes even less profitable. This technology enables a higher economical production in terms of perfectly defined environmental and plant nutrition conditions, resulting in a higher saffron yield and improved quality (Maggio *et al.*, 2006; Fallahi *et al.*, 2017), so that water supply can be reduced particularly in arid areas. Vertical farming technology may also improve the usage of resources, including water, which are always vital in arid conditions. This efficiency in use of resources, especially water, is done through techniques like vertical column, suspended grow bags and plant factories (Touliatos *et al.*, 2016). Controlled environments also allow for increased plant densities, longer flowering periods and superior plant propagation, which in sum result in higher production per unit area and per worker than when planting in the field.

Soilless crop production techniques, hydroponics, in which plant roots are submerged in a nutrient solution either aerated in a static manner or flow-based, and aeroponics, where misted nutrients are delivered, offer alternative methods of saffron field production (Mollafilabi *et al.*, 2013). A number of studies concluded that many advantages exist, where aeroponic and hydroponic saffron corms produced a higher number of flowers and leaves than those produced in a soil medium without affecting stigmas, while also increasing dry weight of corm, as well as the content of crocin and crocetin. Maggio *et al.* (2006) found that saffron can grow and develop in a successful manner in a soilless medium in controlled environments (cold greenhouse and climatic chambers), which showed a higher planting density rate with a 10-day early emergence, as well as earlier flowering in climatic chambers than in a greenhouse. Under these conditions, productivity doubled to around 22 kg/ha, compared to traditional saffron field production, which was 10 kg/ha, leading to the conclusion that using a soilless medium in saffron production seems to be a viable alternative.

Maggio *et al.* (2006) further added that proper use of medium, as well as a controlled environment, would help in increasing flower number as well as accelerating flowering in saffron. However, different studies reported conflicting data concerning the effects of a controlled environment on flower productivity, as it was in traditional saffron field production. There are numerous studies on saffron quality in climatic chambers, which reported conflicting results, where increased saffron quality was found in incubated saffron corms by Poggi *et al.* (2009), while Valle García-Rodríguez *et al.* (2017a) showed the practicability of delayed flower production by storing corms in extremely low oxygen cooling chambers, with a further incubation period, followed by a successful hydroponics, where flowering occurred 2 months later than normal, whereas the same authors reported that saffron quality significantly decreased with prolonged incubation period. In accordance with ISO 3632, the overall extended storage at extremely low oxygen levels with incubation had a negative effect on the total amount of crocin esters and on the picrocrocin value, while the safranal values showed no change. Despite automation being common in other crops, the saffron production process remains based mostly on traditional family farming with limited advancements in growing technology. Although modified planting machinery, like modified onion/potato planting machines, can be employed for planting corms, improper plant head orientation can cause delay in germination and minimize yields. Elevated corm harvesters like potato diggers can help with harvesting corms, but specialized equipment, like modified potato harvesters, other specialized harvesters, modified mechanical harvesters and manual harvesters, can facilitate the harvest of corms only to a limited extent, as in traditional saffron cultivation, specialized equipment for harvesters is not used (Gresta *et al.*, 2008b). Nevertheless, flower harvest, stigma separation and corm growing are carried out entirely by hand due to the fragile flowers, size variation of corms and precise planting techniques (Galigani and Pegna, 1999). Manual farming by family members requires careful cutting of stems with fingernails, performed early in the morning when flowers are closed to obtain optimal stigmas. Destruction of plant leaves around flowers inhibits daughter corm development. This manual farming requires 400 h of labour per kg of dried saffron, with 72% of that being for harvest and separation of stigmas. There seems to be little hope for the automation of this technology, which would result in substantial savings due to reduced labour costs. Technologically advanced greenhouses with automated harvesters, for example, would increase efficiency but decrease total manpower. Smart farming technology, which taps technology from information and communication technology sources like Internet of Things (IoT) sensor technology, geo-position, Big Data and robotics for precision agriculture, can optimize productivity of farms, including small organic ones, to make them efficient and sustainable. Smart farming technology allows farmers to register field variability in a precise manner to use pesticides, fertilizers and other resources in a most effective way and also enhance resource use efficiency, for example by using smaller amounts of water. Nevertheless, acceptance of this technology remains unchanged, with 24% in the EU and 80% in the US in 2011.

There are also traditional ways of saffron farming that involve human labour in countries such as Iran, Kashmir in India, and Spain. Here, *C. sativus* is planted in a well-draining soil type that has a high potential in summer or early autumn. This farming involves the use of conventional farming tools, where growers plant corms in rows that are 20–30 cm apart and 15–20 cm deep (Ghorbani and Koocheki, 2017).

Irrigation, if employed, would be limited to one or two occasions, since saffron prefers a semi-arid climatic condition with a dry summer and cold winter. Weeding and pest control, in traditional practices, would rely on human effort or organic means. Saffron harvest requires an enormous amount of labour, as saffron is harvested manually in the early hours of the morning in order to preserve the quality of the stigmas. Subsequently, stigmas are screened, dried and stored accordingly to maintain their medicinal and fragrance values (Negbi, 1999). With increasing demands for saffron and advancements in scientific research, new technology has been introduced for saffron cultivation that will enhance saffron production, quality and sustainability. A major innovation in this area involves using saffron plantations in raised beds and drip irrigation, which certainly will increase saffron corms' efficiency in using water and maintain their health (Gresta *et al.*, 2008b).

Raised beds would provide better aeration to saffron plantations and avoid compaction of saffron corms, which would be vital for root and corm development. Controlled environment agriculture, including saffron plantations in a greenhouse and soil-hydroponics, has been considered for saffron farming, which would increase saffron production in a controlled environment using coco peat, perlite and other supportive media, which would be a commendable step for continuous saffron yield regardless of climatic conditions. This technology would also result in a decrease in saffron corm contamination due to adverse climatic conditions. Automation and remote sensing technology for saffron farming, inclusive of AI, IoT-based saffron farming, would optimize saffron production and minimize human error in saffron farming. This technology has been in the development stage, but it would be a boon for commercial saffron plantations. The use of other techniques, including plant growth-promoting rhizobacteria and organic plant fertilizers like worm compost and biofertilizers, has proved successful in enhancing saffron flowering, while reducing the use of chemicals in saffron cultivation, which has been an important area of development. Table 3.1 highlights the differences in traditional and modern methods of saffron cultivation and the advantages and limitations of modern methods.

3.3 Traditional and Modern Breeding Approaches

Although Crete's saffron has been recognized as the starting point of plantations, Eurasia, specifically Iran, has a 2000-year-old history of agriculture that might be important for saffron domestication as well (Shokrpour, 2019). Nonetheless, its history of saffron domestication remains unrecognized. Worldwide, saffron has been propagated vegetatively from corms without genetic variation over a long period of time, possibly indicating a universal saffron variety (Mir *et al.*, 2015). Although saffron varieties are clonal, saffron has been cultivated for a long period of time, which would imply that natural mutation, like chromosomal recombination, deletion, inversion, translocation, polyploidy and incomplete and distorted segregation, would introduce genetic variation in saffron (Mir *et al.*, 2015).

In 2023, Iqbal *et al.* published a study in which they identified 272 varieties of elite saffron varieties in terms of their floral, morphological and corm characteristics. In doing so, they found differences in the characterization of saffron flowers and leaves between different plantations.

There are efficient methods of variation in saffron clones despite their presumed genetic identity.

Due to defective meiosis caused by triploidy, hybridization in saffron cannot take place using conventional methods because of sterility. This leads to distorted pollen formation and self-sterility due to defects in sporogenesis/gametogenesis (Chichiriccò, 1999; Caiola, 2004). Although *in vitro* cross-pollination of *C. sativus* female flowers with pollen of related wild taxas *C. cartwrightianus* (Caiola, 1999, 2005), *C. tomasii* (Chichiriccò, 1999) and *C. hadriaticus* (Caiola *et al.*, 2000) produced capsules and fertile seeds, backcrosses were unsuccessful (Caiola, 2005). Evaluation would be necessary for other potential donor taxas (*C. mathewii*, *C. pallasii*, *C. haussknechtii*, *C. almehensis*, *C. michelsonii*) (Alavi-Kia *et al.*, 2008).

Mutation breeding has also been shown to provide an effective alternate means of generating genetic diversity in plant-propagated, sterile crop plants like saffron, where their propagability has aided in tracing, developing and conserving mutants in the M1 generation (Mir *et al.*, 2015) themselves. This approach targets most important parameters like stigma traits, flower/corm production and/or metabolites in stigmas. After targeted mutation with gamma radiation to obtain high-yielding mutants with characteristics

Table 3.1. Comparative analysis of traditional and modern saffron cultivation methods.

Cultivation aspect	Traditional method	Modern method	Advantages	Limitations
Propagation and cultivation	Manual corm planting in open fields	Controlled environment, hydroponics, aeroponics	Precise control, higher yield potential	High capital investment, technical expertise required
Irrigation system	Rain-fed or flood irrigation	Drip irrigation, smart sensors, IoT-based systems	Water conservation, optimal moisture control	Initial setup costs, maintenance complexity
Pest management	Organic methods, manual control	Integrated Pest Management (IPM), biocontrol, intelligent monitoring systems	Reduced crop damage and environmental sustainability	Complex management protocols
Harvesting	Entirely manual at dawn	Semi-mechanized with precision tools	Reduced labour requirements, consistent timing	Stigma separation still requires manual labour
Yield/ha	10–15 kg/ha	≈ 22 kg/ha	Enhanced farmer income, resource efficiency	Technology dependency, skill requirements
Quality control	Variable, weather-dependent	Consistent bioactive compound concentrations	Standardized quality, market reliability	Requires sophisticated monitoring systems

of five-branched stigmas and enhanced dry weight and/or longer dry weight, Khan (2003) found that this mutagen works reliably. Likewise, Khan *et al.* (2011) also reported 0.2 kilorad (KR) gamma radiation as most efficient, which, if standardized, would increase saffron plant development greatly. They used a combination of physical mutagens like gamma rays, along with chemical mutagens like ethyl methanesulphonate (EMS), colchicine and ethidium bromide. Barriers to sterility can also be broken using mutagenesis, as it involves increasing ploidy through a dose of colchicine, as reported in Zaffar *et al.* (2004).

Furthermore, biotechnological advancements offer a powerful arsenal of tools for saffron improvement, as proven by various studies that used molecular markers like RAPD, ISSR, SRAP, SSR and AFLP to estimate genetic diversity in saffron accessions (Rubio-Moraga *et al.*, 2009; Keify and Beiki, 2012; Siracusa *et al.*, 2013; Izadpanah *et al.*, 2014; Shokrpour *et al.*, 2017).

Natural selection plays a key role in plant domestication, and, as saffron grows clonally, one might easily assume that mutations are an important factor in the genetic diversity of saffron. Despite the diversity of saffron varieties, there is no concrete molecular evidence in the form of DNA to support this diversity. Since saffron is triploid, sterile and has no natural ability to produce seeds, it cannot be bred using natural methods, but it can only be propagated vegetatively via its daughter corms, a method which has shown limited success using conventional breeding techniques due to their naturally slow rate of genetic evolution. Due to its naturally slow genetic evolution, conventional breeding techniques have proven to be only moderately successful. This has been further varied due to mutation, deletion, polyploidy and various other genome-related differences in various places over a long period of time, including the varieties 'Lacha' of India and 'Ghyanat' of Iran, and various other cultivars from Italy, Spain and Greece, which differ in yield, quality and other characteristics related to saffron (Salami and Husaini, 2022). To advocate successful plant breeding, it would be important to select superior genotypes, and this would also require evaluation of their genetic diversity, which would form a key factor for their further breeding and conservation for further use as a superior genotype itself, as has been described in Karp *et al.* (1997). Breeding objectives include optimally shaped flowers and corms, high-quality stigmas, a high content of apocarotenoids (especially crocin) and pronounced stress tolerance. Clonal selection with a focus on polyploidy and hybridizations with similar *Crocus* species from the wild, such as *C. cartwrightianus*, *in vitro*, mutagenesis using gamma-radiation, as has been described in various studies by Ahhund-Zade and Muzaferova (1975); Zaffar *et al.* (2004, 2008), and even with gene editing technology by Mir *et al.* (2015), would form potential approaches for their breeding. In spite of saffron being triploid, it has a major drawback due to being a clone, with limited available opportunities for successful development and breeding of new varieties using mutation as a natural means. These concerns are now being aptly met with advancements in saffron's genome, with deep-level genetic analysis available with genotyping genome sequencing, which has been described in Gore *et al.* in 2007. The characterization of genetic diversity, population structure, phylogenetics, the identification of functional genes associated with a specific metabolite and the identification of markers for further breeding progress in various available saffron varieties would also represent a significant advantage of this approach (Soorni *et al.*, 2017). Male sterility due to defective meiosis, with a gametophytic abnormality and defective sporogenesis that creates defective pollen with fertile ovules, has been reported as the major cause of infertility in saffron (Dhar *et al.*, 2017). Male sterility renders saffron incapable of improvement through proper plant reproduction. Therefore, corms are used for vegetative multiplication of saffron, which has greatly restricted plant improvement and saffron production (Sharma *et al.*, 2019). Since a saffron corm that exceeds 8 g weight usually provides only three to four daughter corms, taking a crystal-clear two to three growing periods to flower, this restricted rate of saffron plant multiplication is unavoidable. Fusarium corm rot and rodent attacks also decreased yields, which led to a substantial economic loss (Sharma *et al.*, 2019). Mass multiplication of saffron plantlets through a large-scale tissue culture technique has been deemed essential for saffron plant production, since conventional saffron plant multiplication using corms has proved inefficient (Nehvi and Yasmin, 2017). Although concerns exist about potential mutagenesis that can result from either expression or presence of a transgene, a powerful and parallel approach in plant improvement has been known to be limited due to such concerns. Surprisingly, CRISPR/Cas9-mediated genome editing offers a perfect alternative, as it enables precise manipulation of a target DNA sequence at a specific location in the genome and ensures high precision in the deletion, substitution and insertion of DNA through direct *in vitro* assembly of DNA breaks. This technology has therefore already successfully been used in various plant species, including saffron plants, for different plant improvement strategies, as has been demonstrated in various studies from different laboratories worldwide (Chib *et al.*, 2020).

3.4 Genetic Background and Breeding Challenges

Sterility, which was due to triploidy, and a shortage of genetic variation due to its limited breeding history, make genetic improvement hard because of the large and complex genome represented by *Crocus sativus* (Schmidt *et al.*, 2019; Vahedi *et al.*, 2019). The poor records of breeding culminate in genetic erosion and a lack of superior cultivar types which can be able to adapt to a wide array of environmental conditions or stresses. The success of conserving genetic resources depends on the identification of elite genotypes or clones, which dictates collection, screening and characterization of genetic diversity and population structure (Karp *et al.*, 1997). Regardless of the status of saffron, whether auto- or allotriploid, traditional breeding techniques have had little success in this regard to date. The reason has brought the need to study alternative methods such as protoplast fusion, mutagenesis, chromosome doubling, polyploidy or using wild allies for cross-pollination (Ahooran *et al.*, 2009). More importantly, fertility restoration will make it possible to spread seeds and introduce genetic variability through selection via recombination (Schmidt *et al.*, 2019).

Due to its reliance on high-yielding improved varieties, modern agriculture is intrinsically vulnerable to genetic erosion. This may further result in a loss of valuable genes and genetic resources. This menace is very serious in the case of saffron, which is naturally a sterile triploid plant (Mohammadi and Prasanna, 2003). The strong impact of poor breeding and genetic backgrounds has made saffron highly susceptible to biotic and abiotic stressors through the high level of genetic erosion. Assessing the degree of genetic diversity in plant populations and germplasm collections used in genetic studies and for plant improvement is essential to avoid this disadvantage (Fernández *et al.*, 2011). Genetic diversity can be regarded as an expression of the differences or similarities between species, populations or individuals, which are analysed using molecular methods (Mohammadi and Prasanna, 2003). Knowledge of genetic relationships generally provides important clues for the selection of advantageous genes and traits in breeding programmes aimed at improving individual saffron genotypes.

3.5 Cultivation Practices

3.5.1 Crop establishment

Saffron has specific agronomic requirements in order to achieve maximum yield. It prefers a Mediterranean climate with dry summers and cold winters. An ideal saffron plant grows in a well-draining, slightly acidic to slightly alkaline soil with a pH of 6.0–8.0, as lack of drainage can cause root and corm decay, as reported by Koocheki and Khajeh-Hosseini (2020). Proper site preparation must be accomplished before crop establishment. Tilling to a depth of 25–30 cm using a plough is normal for saffron to break compacted soil and increase root and corm entry into the soil. During site preparation, organic amendments such as farmyard manure/compost (10–15 t/ha) are incorporated into the soil to provide nutrients and amend soil properties, as reported by Gresta *et al.* (2008b). Corm selection and planting techniques are important for successful crop establishment. Healthy corms with no signs of diseases, weighing 8–15 g, are used since larger corms can flower extensively, as reported by Negbi (1999). Corms are usually planted in late summer or early autumn, from August to September in the northern hemisphere, 20–30 cm deep with a distance of 20–30 cm, either in flat beds or in mounds, depending on soil drainage and environmental conditions (Table 3.2).

Saffron has low water requirements since it thrives in semi-arid conditions. However, a shallow watering of 1–2 cm of water can facilitate initial root emergence, and a second watering when flowers are present can increase

Table 3.2. Saffron agricultural calendar and phenological stages.[a]

Time period	Primary activity	Duration	Critical requirements	Optimal conditions
June or early September	Corm planting	2–3 weeks	Healthy corms (8–15 g), disease-free planting material	Planting depth: 15–20 cm, spacing: 20–30 cm
October	Sprouting and initial irrigation	2–3 weeks	Light irrigation to stimulate emergence	Temperature: 15–17°C, adequate soil moisture
November	Flowering and harvesting	2–4 weeks	Dawn harvesting before sunrise	45–55 min/1000 flowers, cool morning temperatures
December–March	Vegetative growth and daughter corm formation	16 weeks	Weed management, limited irrigation	Cool winters, minimal water stress
April–May	Corm maturation	8 weeks	Final irrigation for corm development	Adequate soil moisture, spring temperatures
June–August	Dormancy period	12 weeks	Complete cessation of irrigation	Dry soil conditions, 25–27°C temperature

[a]Phenological calendar based on Mediterranean climate conditions. Timing may vary by 2–4 weeks depending on geographical location and local climate patterns.

stigma production, particularly in arid saffron-growing areas. Waterlogging promotes fungal diseases and root decay. Weeding in saffron plantations has been done manually by people using traditional farming practices. Modern saffron plantations are cultivated using mulching or organic herbicides to manage weeds. According to Chamkhi *et al.* (2023) the use of biofertilizers, worm compost and *Azospirillum*/*Azotobacter* bacterial inoculants has led to an effective increase in flower and saffron yields and a reduction in the use of chemicals. Saffron plantations are usually cultivated for 3–5 years before crop rotation since longer monoculture practices often lead to corm necrosis, increasing saffron attractiveness to various insects. A crop rotation with legumes or cereals is suggested for restoring soil quality and subsequently reducing diseases (Gresta *et al.*, 2016). The plant agricultural calendar and phases are described in Table 3.2.

3.5.2 Influence of soil, climate and irrigation

The ideal soil type for the growth of flowers and the new cormels of saffron is a light, well-drained calcareous clay soil with rich organic matter and slightly alkaline pH. In major growing regions, such soil conditions are seen in Kashmir (high alkalinity), Abruzzo, Italy (medium humus-clay/sandy, high limestone/organic matter, low phosphate), Sardinia (sandy clay alluvial deposits), Spain (friable, lightly calciferous), Greece (sandy clay loam/clay loam, high calcium carbonate content, pH 7.4) and Morocco (sandy/calcareous clay). Thus, this helps in preventing the phenomenon of corm rotting due to waterlogging (Cardone *et al.*, 2020; Koocheki and Seyyedi, 2020; Kothari *et al.*, 2021; Mansotra and Vakhlu, 2022).

Soilless systems such as aeroponics/hydroponics may help crops cope better with changing climatic conditions such as rising temperatures and pest infestations. They can accelerate growth and quality and do not affect the concentration of crocin and crocetin even though the length of roots has been reduced; their effectiveness has always been in debate (Caser *et al.*, 2019; Cardone *et al.*, 2020; Fallahi *et al.*, 2020; Salas *et al.*, 2020). Temperature is crucial for determining phenology. Flower formation begins at summer temperatures of 25–27°C, while cool autumn temperatures at 15–17°C are required for the flowers to appear (Dastranj *et al.*, 2019; Chourak *et al.*, 2021; Wang *et al.*, 2021). However, prolonged incubation at 25–30°C inhibits bud/flower development, and a temperature of ≤15–20°C at the time of sprouting leads to abortion. Best flowering occurs when there is a sequential warm, intermediate and then cool regime. The warm regime should be 30°C for 2–3 weeks; then, for 1–2 months, the temperature should be at 25°C and, finally, cool, below 15–20°C by early October (Wang *et al.*, 2021).

Irrigation practices vary considerably between individual saffron-growing regions. In Greece, Sardinia and Abruzzo, for example, no additional irrigation is required due to sufficient rainfall in autumn. Contrarily, 70% of the cultivated area is irrigated in Castile-La Mancha, Spain. Morocco applies 350–500 m^3/ha weekly from September to November, and every 2 weeks during December to March. Irrigation is completely stopped during the dormant period of the corms, which lasts from April to August (Table 3.3).

In Kashmir, India, low rainfall and insufficient irrigation lead to significant production losses (Husaini *et al.*, 2010). Projections indicate that the average annual water requirement for saffron will rise to 490 mm by 2040, attributed to decreasing precipitation (Jafarzadeh *et al.*, 2015). This is consistent with reported potential evapotranspiration values of 523 mm and 640 mm for successive growing seasons (Yarami *et al.*, 2011). Given the anticipated scarcity of water resources, enhancing irrigation water use efficiency is essential to meet future crop demands. Research by Mollafilabi *et al.* (2020) demonstrates that both maximum stigma dry weight and water use efficiency were achieved in the third growing season under drip irrigation, establishing it as the preferred method for production. While saffron quality remained acceptable under drip irrigation, the sprinkler method yielded a marginally superior product due to a higher crocin content.

3.5.3 Pest and disease management

Despite saffron being a valuable spice with many uses in traditional medicine and cuisine, saffron production has been steadily declining over the years. The factor that most affects successful saffron production as a crop is that of saffron corms that are needed for planting being in inadequate conditions (Gupta *et al.*, 2021). The extent of saffron being cultivated has greatly decreased globally owing to a number of biotic as well as abiotic conditions. Corm root rot has been recognized as a major factor that suppresses successful saffron crop development in both established and developing areas. The bulb mite, known as *Rhizoglyphus robini*, represents a major challenge to saffron production. This mite burrows into saffron corms, which in turn results in dark spots, saffron root rot and a shortage of saffron in a saffron harvest. This condition, coupled with other symptoms that entail lemony-yellow saffron plant leaves, saffron stigmas being shorter than normal and a low saffron crop yield, has been of major concern among saffron producers, as reported by Rahimi and Kamali in 1993. To counter this challenge, it has been advocated that saffron producers avoid irrigating their saffron crops in summer, dipping saffron corms in acaricides before planting, planting saffron corms 18–20 cm deep in well-draining soils with additional sand for heavy soils, as well as performing soil solarization for 30 days, as stipulated in the studies of Rahimi *et al.* (2018).

Rodents, in particular the short-tailed bandicoot rat (*Nesokia indica*) and the southern mole vole (*Ellobius fuscocapillus*), are responsible for significant damage to saffron due to their feeding on corms and tunnelling, which spills irrigation water (Shahrokhi *et al.*, 2006). Control methods for rodent infestation in saffron farms are injecting smoke, such as that from motorcycles and sulfur, into holes and tunnels, applying season-specific toxic baits, which are nutrient-dense in winter and juicy in summer, and ploughing along the

Table 3.3. Suitable climatic conditions and soil requirements for saffron cultivation.

Parameter	Optimal range	Tolerance range	Critical notes
Soil type	Sandy loam to clay loam	Well-drained calcareous soils	Avoid waterlogged conditions
Soil pH	6.0–8.0 (neutral to alkaline)	6.5–7.5 optimal	High alkalinity tolerated
Drainage	Excellent	Good to excellent	Poor drainage causes corm rot
Organic matter	15–20 t/ha	8–50 t/ha	Improves soil structure
Altitude	800–1800 m	600–2000 m	Higher altitudes preferred
Annual rainfall	300–500 mm	250–600 mm	Concentrated in autumn/spring
Summer temperature	25–27°C	20–30°C	Required for flower initiation
Autumn temperature	15–17°C	12–20°C	Critical for flower emergence
Winter temperature	5–15°C	Up to –22°C	Cold tolerance essential
Humidity	Low to moderate	<70% relative	High humidity promotes disease
Sunshine hours	High intensity	Full sun exposure	Affects stigma quality
Frost tolerance	Moderate	Can survive light frost	Protect during severe frost

perimeter to disturb nests. Fungal pathogens like *Fusarium solani, F. oxysporum* and *Sclerotinia rolfsii* cause corm rot, with symptoms of dark brown lesions, powdery soft rot and leaf chlorosis (Ghani, 2002; Ahmad and Sagar, 2007; Kalha *et al.*, 2007). This can be managed by immersing corms in 0.2% Tecto (thiabendazole) or Bavistin (carbendazim) for 30 min (Sud *et al.*, 1999). Bacterial soft rot, caused by *Burkholderia gladioli*, infects buds and leaves before flower initiation, forming lesions, reducing flower production and resulting in flower vandalism (Fiori *et al.*, 2011). The bean yellow mosaic virus and turnip mosaic virus are other viruses that affect saffron but are not common in all saffron-growing areas. These viruses are usually found in other nations, like Italy, as reported by Russo *et al.* in 1979 for bean yellow mosaic virus, and in 2000 by Chen for turnip mosaic viruses. They can be managed through vector management and other biotechnologies like meristem culture and thermotherapy. When controlling pests in saffron cultivation, it is generally very important to take preventive measures. These include growing saffron using disease-free corms, not growing saffron in waterlogged soils (Cardone *et al.*, 2020) and carrying out a 6-week period of solarized soils to eliminate saffron pathogens as well as weed seeds.

3.6 Harvesting and Post-harvesting Processing

To maintain saffron of superior quality, its flowers are gathered manually in early morning hours, requiring 45–55 min for every 1000 flowers. Separation of stigmas would take over 100–130 min, amounting to 370–470 h/kg of dried saffron (Cardone *et al.*, 2020). Saffron flowers are to be gathered immediately upon flowering, when stigmas are deep red in vibrant colour. Today, there are mechanical harvesting machines that remove the corolla using negative pressure. However, saffron flowers are mainly separated from the stigmas by hand. To maintain quality, only 2 mm of the style may remain from the tips. To preserve saffron properties, drying of saffron plays a vital role, resulting in an 80% weight decrease for stigmas. Thus, drying of saffron has a profoundly major consequence on saffron aroma and flavour too, with reduced weight of stigmas. Different drying methods are used in the countries where saffron is grown: in Spain, the stigmas are roasted at 70°C for 30 min, and in Greece, they are gradually heated from 20°C to 30–35°C in the dark, which takes about 12–24 h. In Morocco, on the other hand, the stigmas are dried in the sun for about 2 h or in the shade for 7–10 days. In Kashmir, the stigmas are also dried in the shade, but for 27–53 h. The completely dried saffron threads, with a moisture content of about 8–10%, are finally packed in dark, completely airtight bags and protected from direct sunlight.

3.7 Conclusion

Researchers are striving to improve both the quality and quantity of saffron. However, saffron is naturally sterile and has low genetic variability, which precludes the use of conventional breeding methods. A long history of growing, especially in certain areas such as Iran, indicates that natural mutation can occur from time to time. This also raises the possibility of using clonal selection for increasing yields in saffron. Mutation breeding, genetic engineering, *in vitro* techniques, as well as crosses of interspecific varieties, are considered necessary in order to introduce new genetic variations. Despite all efforts to introduce genetic variability and find suitable clones, research into genetic improvement is still in its early stages, and no superior varieties have been bred to date. In view of this worldwide homogeneity of saffron as a single variety, with insufficient improvement in yields through a search for superior clones owing to lack of sufficient natural variation, it appears that introducing genetic variation through techniques such as crosses in interspecific varieties, mutagenesis and genetic engineering is becoming a prerequisite for further improvement.

The extent of reliance on vegetative propagation using corms also impedes mass production, susceptibility to soil-borne diseases and labour costs, particularly in harvesting and post-harvest processing. New ways of plant production, including soilless farming techniques like hydroponics and aeroponic systems, controlled environments and smart farming, provide a bright outlook for enhanced yields, quality and efficient use of resources. However, their use for saffron farming remains prohibitive due to their relatively high initial costs. The use of advanced biotechnologies, including mutagenesis, cross-breeding with wild *Crocus* species and genome editing using CRISPR/Cas9, has been found fundamental in saffron farming for generating additional valuable traits that are not otherwise present in this crop due to sterility. Further enhancement in automation technology for harvesters, generating drought-tolerant varieties and adopting IoT-based precision farming practices will be vital for sustainable intensification, particularly in a changing climatic scenario of decreased water availability. International cooperation in research, facilitated by genome and omics-based approaches, has been found fundamental in realizing the genetic potential of saffron, maintaining a stable supply chain and mitigating increasing global demands, as well as ameliorating small-scale saffron farmers in traditional and new saffron-producing areas.

4 Chemistry: Bioactive Substances and Analytical Characterization of Saffron Extracts

Hassan Rezadoost and Hartwig Schulz

4.1 Introduction

According to World Flora Online (WFO) and Plants of the World Online (POWO), over 200 different species of the genus *Crocus* have been described, of which the following eight species are discussed in more detail in this chapter: *Crocus almehensis*, *Crocus michelsonii*, *Crocus biflorus*, *Crocus caspius*, *Colchicum speciosum*, *Crocus speciosus*, *Crocus cancellatus* and *Crocus haussknechtii* (Fig. 4.1).

4.2 Chemical Composition

Saffron contains several secondary substances (Fig. 4.2) that can be assigned to different chemical compound classes. In terms of its commercial use, its characteristic colour and taste are particularly important. In addition to its use in cooking, saffron is also very popular as a medicine due to its various pharmacological effects, which are mainly attributable to its individual bioactive components. It is therefore very important to carefully analyse its chemical composition, as this is not only helpful for testing its purity, but also important for optimizing its various uses. The most important ingredients include carotenoids, which also include numerous apocarotenoid derivatives that are responsible for the individual colour of saffron threads (Ahrazem *et al.*, 2016). Among these, crocins (the glycosylated esters of crocin) play a special role in the yellow-orange colouring due to their water solubility, whereas crocin as the aglycone component in crocins is responsible for the above-mentioned property. Safranal, on the other hand, makes an important contribution to the characteristic aroma of saffron. In addition, various alcohols, aldehydes, ketones and terpenes also play an important role in the formation of smell and taste (Demurtas *et al.*, 2018).

Fig. 4.1. Various species of the genus *Crocus*: (A) *C. biflorus*, (B) *C. michelsonii*, (C) *C. caspius*, (D) *C. speciosus*, (E) *C. cancellatus*, (F) *C. almehensis* (https://www.worldplants.de/world-plants-complete-list/complete-plant-list/?name=Crocus-longiflorus#plantUId-75331), (G) *C. haussknechtii* (https://www.inaturalist.org/taxa/1584369-Crocus-haussknechtii) and (H) *Colchicum speciosum* (https://powo.science.kew.org/taxon/urn:lsid:ipni.org:names:533389-1).

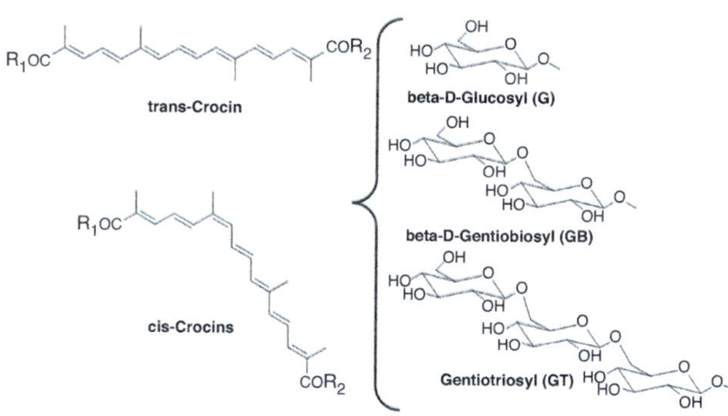

Compound	R1	R2
Crocetin	H	H
trans-crocin 1	H	G
trans-crocin 2	H	GB
cis-crocin 2	H	GB
trans-crocin 3	G	GB
cis-crocin 3	G	GB
trans-crocin 4	GB	GB
cis-crocin 4	GB	GB
trans-crocin 5	GB	GT
trans-crocin 6	GT	GT

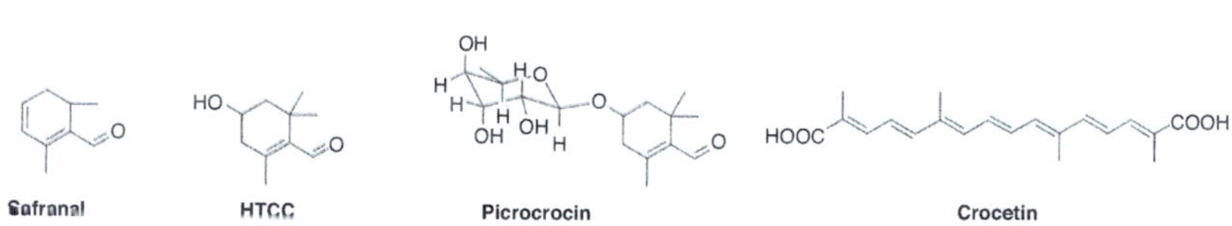

Fig. 4.2. Most important chemical substances of saffron.

4.3 Major Bioactive Substances

4.3.1 Crocins

Crocins are water-soluble carotenoids responsible for the deep red colour of saffron (Demurtas *et al.*, 2018). They are esters/diesters of crocetin with sugars such as gentiotriose (O-β-D-glucopyranosyl-(1→6)-O-[β-D-glucopyranosyl-(1→6)]-D-glucose, GT), gentiobiose (β-D-glucopyranosyl-(1→6)-D-glucose, GB) and glucose (G), existing in different geometric isomeric forms, with the *trans* isomers predominating over the *cis* forms (Fig. 4.3).

The most common form, *trans*-crocin-4 (also known as crocin-1 or α-crocin), contributes significantly to the colouring power of saffron. The content of this compound is usually measured photometrically at 440 nm in a 1% (w/v) solution. Crocins are unique among carotenoids in that, as hydrophilic substances, they produce an intense yellow colour in aqueous solutions at concentrations as low as 1 mg/kg. Crocins account for 6–16% of saffron's dry matter and are known for their potent antioxidant properties, capable of quenching free radicals and protecting cells and tissues from oxidative damage. They exhibit additional bioactivities, including neuroprotective, anti-inflammatory and anti-cancer effects. Crocins are also stable at high temperatures, enhancing their versatility in culinary applications; however, they are sensitive to light-induced oxidation, which can lead to fading of their colour (López-Jiménez *et al.*, 2021). The most common forms of crocins found in saffron differ primarily in the number, type and position of the sugar units esterified to the crocetin backbone, as follows: crocin-6 (hexaglycosylated crocetin), crocin-5 (pentaglycosylated crocetin), crocetin-4 (tetraglycosylated crocetin), crocetin-3 (triglycosylated crocetin), crocetin-2 (diglycosylated crocetin), crocetin-1 (monoglycosylated crocetin), β-D-glycosylated crocetin methyl ester and gentiobiosyl crocetin methyl ester. Among these, the most common form is *trans*-crocin-4, which exhibits an intense signal in the HPLC-DAD chromatogramme. Si *et al.* (2022) developed a combined targeted and untargeted UPLC-ESI/MS approach for the comprehensive characterization of the chemical constituents of saffron (*Crocus sativus* L.). The analysis of 21 saffron samples from six regions led to the identification of 47 crocins (including 32 previously unknown compounds and 6 with novel structures) and 91 non-crocins (including 43 previously unknown compounds) (Fig. 4.3) (Si *et al.*, 2022).

4.3.2 Picrocrocin

Picrocrocin is a monoterpene glycoside responsible for the peculiar bitter taste of saffron (Figs. 4.2 and 4.3). It is a specific degradation product of zeaxanthin and represents a direct precursor of safranal, responsible for the aroma of saffron. During the drying process, picrocrocin is converted into safranal through thermal and enzymatic reactions. Picrocrocin is found exclusively in species of the genus *Crocus* and is an important marker for assessing the quality and authenticity of saffron. It usually accounts for 1–13% of the dry

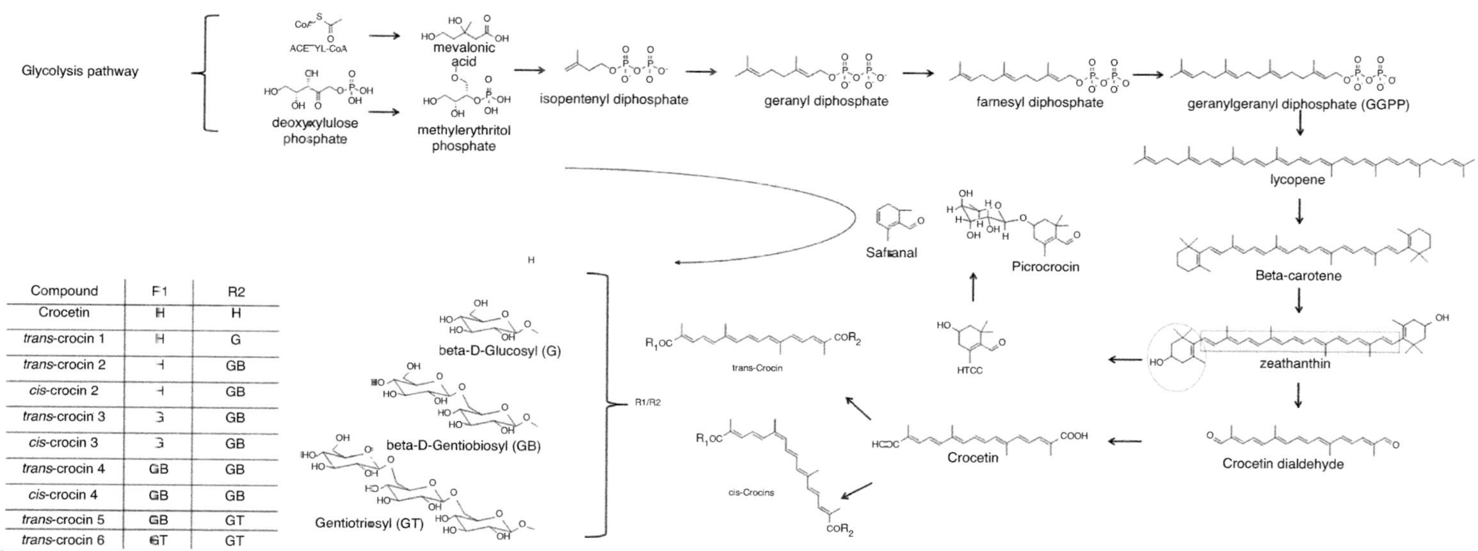

Compound	F1	R2
Crocetin	H	H
trans-crocin 1	H	G
trans-crocin 2	H	GB
cis-crocin 2	H	GB
trans-crocin 3	G	GB
cis-crocin 3	G	GB
trans-crocin 4	GB	GB
cis-crocin 4	GB	GB
trans-crocin 5	GB	GT
trans-crocin 6	GT	GT

Fig. 4.3. Most important saffron substances and their biosynthesis pathways.

weight of saffron and is also said to have a protective effect by deterring insects from eating the plant's stigmas (Diretto *et al.*, 2019).

4.3.3 Safranal

The most important volatile component of saffron, safranal, is produced by deglycosylation and further degradation of picrocrocin during the drying and storage of saffron threads. This process involves thermal and enzymatic reactions. As a monoterpenoid aldehyde, safranal is mainly responsible for the characteristic hay-like smell of saffron. In addition to its function as a flavouring agent, safranal is also said to have antioxidant, anti-spasmodic and antidepressant properties. It is usually analysed and studied using techniques such as GC-MS and GC-headspace. The hydrodistillation process yields an essential oil content of about 0.5% (w/w), of which about 75% is safranal, making it the main component of saffron essential oil (Ahrazem *et al.*, 2016; Cid-Pérez *et al.*, 2021).

4.3.4 Minor constituents and volatiles

In addition to the most important bioactive compounds, saffron stigmas contain a variety of further volatile and non-volatile components, including flavonoids (especially kaempferol derivatives), terpenoids and various aldehydes and esters. These secondary substances are thought to contribute synergistically to the biological activity of saffron. In addition, they are used as important markers in determining the geographical origin and identification of different saffron genotypes. Besides the well-known stigma, saffron leaves are also a valuable source of bioactive phytopharmaceuticals, including flavonols such as quercetin, kaempferol and galangin, and anthocyanins such as delphinidin. Among these, kaempferol has an excellent antioxidant profile and pharmacokinetic profile compared to quercetin. Surprisingly, saffron leaves contain more than 100 times the amount of kaempferol than commonly accepted food sources with remarkable amounts of iron, which is why this material is referred to as a nutritious and extremely valuable by-product of saffron processing (Demurtas *et al.*, 2019).

4.4 Saffron Biosynthesis Pathway

The characteristic metabolites of saffron, crocin, picrocrocin and safranal, can be considered as final products of a stigma-specific apocarotenoid synthesis. While stigmas primarily metabolize crocins and picrocrocin, safranal is mainly metabolized after harvesting. Their common biochemical basis can be tracked back to the plastidial carotenoid fraction, more precisely on zeaxanthin, a substrate that is cleaved by a CCD enzyme specific to saffron. All these facts and ideas have been investigated and documented using various methods in ISI-indexed publications (Moise *et al.*, 2014).

4.4.1 Production of carotenoid precursors in plastids: isoprenoids → zeaxanthin

It begins with the plastid biosynthesis of C5 isoprenoids, IPP/DMAPP (Isopentenyl PyroPhosphate/Dimethylallyl PyroPhosphate), via the MEP (Methylerythritol 4-phosphate) metabolic pathway as a GGPP (Geranylgeranyl pyrophosphate) substrate source. GGPP is a direct precursor of carotenoids. In plants, there is a highly conserved biosynthesis sequence from GGPP substrates. The enzyme phytoene synthase (PSY) converts GGPP to phytoene. Further desaturations and isomerizations take place to finally form lycopene from phytoene. The cyclization of lycopene then leads to β-carotene, which is then hydroxylated at specific sites to form xanthophylls, in particular zeaxanthin. All these steps are documented in detail in review articles on carotenoids (Walter and Strack, 2011; Moise *et al.*, 2014; Nisar *et al.*, 2015). Within *Crocus sativus* stigmas, transitions involving developmental changes in plastid type (closer to chromoplasts) occur with a prominent biosynthesis of carotenoids and transcription of apocarotenoids biosynthesis genes.

The salient feature with implications for saffron obtained from these plants regards the role whereby zeaxanthin becomes the substrate for a committed step branching into crocins and picrocrocin production (Ahrazem *et al.*, 2016; Demurtas *et al.*, 2018).

4.4.2 The dedicated 'signature' cleavage reaction: zeaxanthin → crocetin dialdehyde + HTCC

The main reaction characteristic of apocarotenoid biosynthesis in saffron is mediated by a carotenoid cleavage dioxygenase (CsCCD2) and shown to be responsible for zeaxanthin conversion into the direct apocarotenoid precursors of the main metabolites of saffron (Ahrazem *et al.*, 2016; Demurtas *et al.*, 2018). CsCCD2 acts on zeaxanthin at the 7,8 and 7′,8′ double bonds, cleaving it into crocetin dialdehyde (C20) and the C10 aldehyde (2,6,6-trimethyl-4-hydroxy-1-cyclohexene-1-carboxaldehyde). Plastid association occurs for this reaction due to zeaxanthin membrane association and localization within plastids/chromoplasts (Ahrazem *et al.*, 2016; Demurtas *et al.*, 2018). It thus represents the main reason for an essential biological attribute of saffron: colour components (crocins) and bitterness/aroma precursors (picrocrocin/safranal) are biochemically coupled, because they originate from the same carotenoid substrate through a shared committed reaction (Ahrazem *et al.*, 2016; Demurtas *et al.*, 2018).

4.4.3 Oxidation with formation of the crocin aglycone: crocetin dialdehyde → crocetin

Crocins are glycosylated derivatives of crocetin and not crocetin dialdehyde. Consequently, the dialdehyde derived from CsCCD2 will have to be oxidized to its respective dicarboxylic acid. Indeed, this is the second committed step for crocin biosynthesis in which it establishes the apocarotenoid structure and readies it for glycosylation (Demurtas *et al.*, 2018).

4.4.4 Glycosylation and diversification: crocetin → crocins

The deep red colour of saffron arises from pigments known as crocins, which are a series of crocetin glycosyl esters with varying numbers and arrangements of glucose moieties. Glycosylation is essential, as crocetin itself is fairly hydrophobic; nonetheless, crocins are highly water-soluble and can be concentrated very highly within the stigma (Demurtas *et al.*, 2016). A very important enzyme that plays a role in the early synthesis of crocin molecules is CsUGT74AD1, a UDP-glycosyltransferase enzyme that has been shown to be involved in the conversion of crocetin into crocins with either one sugar unit or two sugar units (designated as crocin 1 and 2' within the enzymatic system adopted in that research) (Demurtas *et al.*, 2016). This initial glycosylation step 'preps' crocetin for subsequent sugar addition, facilitating the synthesis of more highly glycosylated crocins, which are prominent molecules within. Moreover, for very glucosylated molecules like crocins, there is evidence for a secondary glycosylation reaction driven by UGT91P3. A subsequent study conducted by López-Jiménez *et al.* (2021) suggested that the enzyme UGT91P3 primarily produces crocins with a chain length of four to five glucose moieties, and that the combination of UGT91P3 with CCD2 and UGT74AD1 enzyme activities within an expression system strengthens the role of UGT91P3 as an essential 'finishing enzyme' for crocin (López-Jiménez *et al.*, 2021). All these results combine in support of a multi-stage assembly line as follows: (i) production of crocetin dialdehyde by CsCCD2; (ii) production of crocetin by CsALDH3I1; (iii) introduction of the first glucose molecules by CsUGT74AD1; and (iv) contribution to the synthesis of highly glucosylated crocin.

4.4.5 Parallel branch to bitterness: HTCC → picrocrocin

Simultaneously with crocin biosynthesis, CsCCD2 is also responsible for the production of HTCC, the C10 aldehyde that constitutes picrocrocin, the main bitter compound of saffron and thus far its most widely recognized precursor for safranal. The stabilization of HTCC as picrocrocin occurs via glucosylation. UGT709G1, a UDP-glycosyltransferase responsible for picrocrocin biosynthesis in saffron, was identified and characterized. Accordingly, there is enzymatic understanding on how HTCC is 'trapped and reserved' within the plant as the glycoside (Diretto *et al.*, 2019). It represents a branching structure which has conceptual significance, as it highlights that glycosylation plays a role within saffron's biology as follows: (i) making crocetin soluble and stable as crocins and (ii) efficiently storing a volatile compound's aldehyde form as picrocrocin as a glycoside (Demurtas *et al.*, 2018; Diretto *et al.*, 2019).

4.4.6 Aroma generation during processing: picrocrocin → safranal

Contrary to crocin and picrocrocin, the amount of safranal is generally low or zero before drying and becomes detectable largely after drying and storage. Safranal preferentially depends on storage as an end product of the living stigma metabolite pathway. From volatile compound analysis of saffron, it becomes clear that the amount of safranal varies greatly with drying and storage (Maggi *et al.*, 2010; Cid-Pérez *et al.*, 2021). The most accepted biochemical reaction for picrocrocin involved deglycosylation processes (either enzymatic and/or non-enzymatic, depending on the conditions) followed by dehydration/rearrangement reactions to form safranal. Higher temperatures and times may hasten these reactions, but these may potentially result in the deterioration of crocins and overall product quality (Sánchez *et al.*, 2011; Cid-Pérez *et al.*, 2021). Kinetic experiments have clearly shown that picrocrocin decomposition is temperature-dependent and thus substantiates the large influence of processing variables on saffron aroma (Sánchez *et al.*, 2011). Consequently, a full description of a 'biosynthesis pathway' for saffron compounds would have to include as steps (i) the enzymatic reactions involved in crocin and picrocrocin production within the plant and (ii) picrocrocin conversion to safranal (Maggi *et al.*, 2010; Sánchez *et al.*, 2011; Diretto *et al.*, 2019; Cid-Pérez *et al.*, 2021).

4.4.7 Transport and storage: vacuolar sequestration of crocins via ABCC transporters

Safranal and crocin accumulate in very high amounts in saffron and are eventually deposited into a vacuole. To be deposited into a vacuole, active transport against concentration gradients is needed. A strategy using co-expression and transport function screening led researchers to discover the involvement of ABCC transporters in vacuolar accumulation of crocin in saffron stigmas and offered proof for what might be described as the 'logistics' phase of crocin biology (Demurtas *et al.*, 2019). However, what appears here as transport is not a trivial process, as vacuolar sequestration contributes significantly to providing a strong metabolic sink. It thus aids in preventing any disruption of cytosolic physiology due to excessive levels of pigments. It should thus be conceptualized as an integrated unit with functions like biosynthesis, modification, compartmentation and transport (Demurtas *et al.*, 2018, 2019).

4.5 Enzyme Incorporated to Saffron Biosynthesis

The characteristic metabolites of saffron (*Crocus sativus* L.), crocins (pigments), picrocrocin (bitterness) and safranal (aroma), are apocarotenoids which stem from the plastidial carotenoid pathway and then undergo stigma-specialized cleavage and tailoring reactions, so that the enzymes involved can be grouped into four functional classes: (i) precursor-supplying and carotenoid-forming enzymes; (ii) apocarotenoid 'commitment' enzymes; (iii) tailoring enzymes that convert and decorate apocarotenoids; and (iv) processing-related enzymes and accessory proteins enabling accumulation and aroma development (International Organization for Standardization [ISO], 2010). In the upstream class, plastids provide the carotenoid substrate pool – particularly zeaxanthin – by the conserved plant sequence from isoprenoid precursors via phytoene and then by desaturation/isomerization to lycopene, further cyclization/hydroxylation yielding xanthophylls like zeaxanthin (Sánchez *et al.*, 2011; Daruwalla and Kiser, 2020); in saffron stigmas, transcriptomic data point to the enrichment of early and mid-carotenoid genes – including PSY and PDS, desaturation/isomerization components such as ZISO, as well as key downstream flux-control steps such as LCYB and BCH – consistent with active channelling of metabolic flux towards zeaxanthin into the tissue where apocarotenoids build up (Walter and Strack, 2011). Within this upstream set, phytoene synthase is particularly relevant because it catalyses the first committed carotenoid step; in *C. sativus*, multiple PSY genes show tissue- and stage-specific expression; notably, CsPSY2 is predominantly detected in stigmas and displays a temporal pattern parallel to the accumulation of crocins, suggesting that saffron partly controls apocarotenoid output at the level of carotenoid entry flux by regulating the PSY step (Moise *et al.*, 2014). Overall, representative upstream enzymes include PSY, PDS, the isomerization/desaturation network, LCYB, BCH and other carotenoid pathway components jointly building the pool of zeaxanthin (Sánchez *et al.*, 2011; Walter and Strack, 2011; Moise *et al.*, 2014; Daruwalla and Kiser, 2020). Further downstream, the pathway enters the hallmark specialization of saffron through the 'commitment' class, in which the key reaction is the signature cleavage of zeaxanthin forming the first committed apocarotenoid intermediates. After cleavage, saffron depends on the tailoring class – dominating oxidation and glycosylation – to stabilize the products, increase their solubility and enable massive accumulation. Because crocins are glycosylated esters of crocetin – a dicarboxylic acid – rather than of crocetin dialdehyde, an oxidoreductive step is necessary, and Demurtas and colleagues identified CsALDH3I1, an aldehyde dehydrogenase that efficiently converts crocetin dialdehyde into crocetin and whose characterization supports a compartmentalized pathway with membrane-associated features rather than a single-organelle process. Glycosylation is then responsible for both pigment solubility and precursor stabilization: the first characterized crocin-forming glycosylation in saffron is catalysed by CsUGT74AD1, a UDP-glycosyltransferase that converts crocetin into initial forms of crocin – often described, in that experimental framework, as crocins 1 and 2′ – while highly glucosylated crocins, typical of saffron quality, involve later-stage glycosylation by UGT91P3, which *in vitro* predominantly produces crocins with four and five glucose residues and, in combination with CCD2 and UGT74AD1 in transient expression experiments, supports its role as a key final-step glucosyltransferase in the crocin enzyme set. In parallel, the bitterness compound picrocrocin is produced through the glycosylation of a cleavage-derived aldehydic intermediate to a non-volatile glycoside, and UGT709G1 has been identified as a UDP-glycosyltransferase involved in picrocrocin biosynthesis, explicitly linking it to the branch that ultimately yields safranal during processing. Finally, the processing-related and accessory class captures that safranal is largely associated with post-harvest drying and storage, whereby picrocrocin can be converted into safranal through the combined action of heat and β-glucosidase activity, with enzymatic hydrolysis and non-enzymatic thermal transformation likely acting in concert depending on the processing conditions, and that high crocin accumulation depends on pathway-enabling logistics, with vacuolar sequestration being crucial and tonoplast transport machinery, including ABCC transporters, implicated in mediating vacuolar crocin accumulation in saffron stigmas and integrating biosynthesis with storage capacity.

The pathway of this enzyme may also be similar in other plants. The results highlighted that successful biosynthesis of saffron pigments, including crocins, in *Nicotiana benthamiana* depends on a harmonized expression of four key enzymes from *Gardenia jasminoides*, which includes carotenoid-cleaving dioxygenase GjCCD4a, aldehyde dehydrogenase GjALDH2C3, and two UDP-glycosyltransferases, GjUGT74F8 and GjUGT94E13. Collectively, these enzymes reconstruct the full crocin pathway: CCD4a cleaves the carotenoid substrates into producing crocetin dialdehyde, ALDH2C3 oxidizes to form crocetin, and the two UGTs sequentially glycosylate the crocetin into producing crocins. Genetic analyses revealed strong transgene expression in T0 and T1 plants, with specific expression patterns shaping the metabolite profile: higher GjUGT94E13 favoured the production of crocin I, while higher GjUGT74F8 favoured crocin II. Enzymatically, introduction of these UGTs increased the downstream conversion efficiency manyfold, with the result that crocetin and lower crocins (III–V) were absent (Fig. 4.4).

This resulted in crocin I and II reaching up to 99% of total crocins. Overall, the multigene stacking strategy evidenced that enzyme choice and relative expression balances directly determine the crocin composition as a means of providing a genetically stable and efficient platform for the production of saffron-like metabolites in planta (ISO, 2010).

4.6 Chemical Profiling Techniques

Because of the rising importance of saffron and its propensity for contamination and deterioration, authenticity and safety have instead become interwoven challenges. As authentication implies, substitutions with similar botanicals, introduction of dyes, mixing with discarded stigmas and misleading geographical origins are problems that communicate mere market deception. However, they can concurrently introduce safety challenges with unforeseen levels and concentrations of added chemicals. Simultaneously, safety considerations extend beyond deliberate deception via contamination and deterioration driven by considerations involving harvest methods, dehydration, storage and packaging conditions. By extension, these variables impact and potentially drive changes in defining factors and trace components associated with safety. Regarding emerging methods suited to resolve these challenges, emerging integrations involving spectral imaging and artificial intelligence, and more specifically Artificial

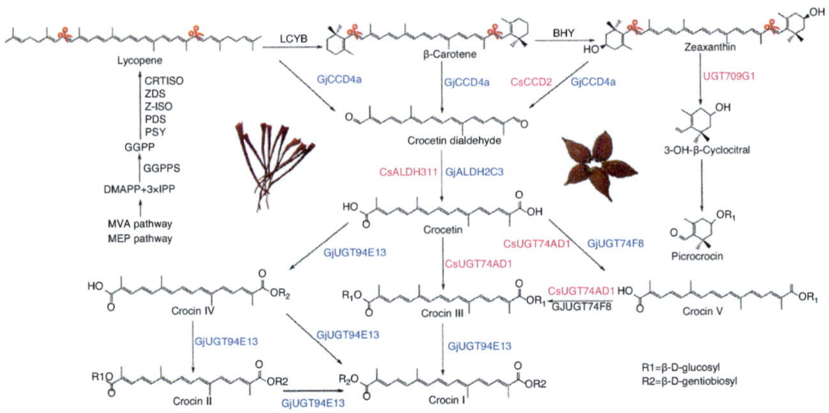

Fig. 4.4. The crocin biosynthetic pathway in *Crocus sativus* L. (pink) and *Gardenia jasminoides* (blue). PSY, phytoene synthase; PDS, phytoene desaturase; Z-ISO, ζ-carotene isomerase; ZDS, ζ-carotene desaturase; CRTISO, carotenoid isomerase; LCYB, lycopene β-cyclase; BHY, β-carotene hydrolase; GjCCD4a and CsCCD2, carotenoid cleavage dioxygenase; GjALDH and CsALDH3I1, aldehyde dehydrogenase; GjUGT74F8, GjUGT94E13, CsUGT74AD1 and UGT709G1, UDP-glucosyltransferase (Xie *et al.*, 2023). Figure used under licence CC-BY-4.0.

Intelligence-Hyperspectral Imaging Analysis (AI-HSI), stand as highly attractive due to their immediate capabilities for marrying detailed spectral pattern recognition with cognitive intelligence suited for simultaneous authentication and determination of defining attributes against genuine versus adulterated and lower-quality samples. Going beyond traditional authentication profiles, thus, intersectionality involving spectral imaging and AI-HSI announces an exciting new frontier for rapid and reliable authentication decisions.

4.6.1 UV-visible spectrophotometry

UV-Vis spectrophotometry represents the widely accepted validated screening technique for saffron quality because it directly relates to international standardization and historical Colouring Strength (CS) practice. The ISO 3632 system institutionalized absorbance indices via UV-Vis as approximate indicators for three attributes, which define the value for perception of saffron: colouring power, dominated by crocin/crocetin esters, taste, associated with picrocrocin, and aroma-related capabilities, which refer to absorbance characteristics related to safranal. The ISO 3632 standard is the most widely adopted method for assessing the quality and classification of saffron stigmas. This method relies on ultraviolet-visible (UV-Vis) spectrophotometry to measure the absorbance of saffron aqueous extracts at specific wavelengths corresponding to its principal bioactive components: crocins at 440 nm, picrocrocin at 257 nm and safranal at 330 nm. Based on these absorbance values, saffron is categorized into three distinct quality grades: Grade I (picrocrocin ≥70, crocin

≥200, safranal 20–50), Grade II (picrocrocin ≥55, crocin ≥170, safranal 20–50) and Grade III (picrocrocin ≥40, crocin ≥120, safranal 20–50). While the ISO 3632 method is widely accepted due to its simplicity, cost-effectiveness and ease of application, it is not without limitations. One of the primary concerns is spectral interference among the compounds, particularly the overlap of crocins with the absorbance of picrocrocin and safranal. Because crocins exhibit strong absorbance in the UV region (Fig. 4.5), their presence at high concentrations can lead to an overestimation of picrocrocin and safranal levels (Kabiri *et al.*, 2017). This overlap compromises the specificity and accuracy of the method, especially in high-quality saffron samples rich in crocins. Therefore, while ISO 3632 serves as a valuable preliminary screening tool, complementary analytical techniques may be necessary for precise quantification and comprehensive quality assessment.

Quality control procedures incorporate this as a first-pass screen for: (i) sample grading; (ii) measurement of veining loss due to ageing; and (iii) indication of samples with atypical gross dilution. The literature also explains the reasons for suppressors and rogue suppliers not considering authenticating with conventional methods, and what role UV-Vis would have within these considerations: saffron contains more than one cumulative chromophore, and more so within sample crocin isomers. Conditions exist wherein samples with different chemistry would have similar index values (Orfanou and Tsimidou, 1996).

A very relevant validation finding is that aroma calculations based on UV-Vis spectra may differ from compound-specific analyses, as no/weak correlations were found for safranal surrogates obtained via UV-Vis spectroscopy and quantification as well as systematic overestimation due to other co-exTRACTs contributing to absorbance, which plays a crucial role regarding authenticity assertions encompassing processing, ageing and addition of colourants (del Campo *et al.*, 2010).

Due to these limitations, it is necessary that today's best practice employs UV-Vis as a triage tool: it offers rapid, normalized comparability with particular

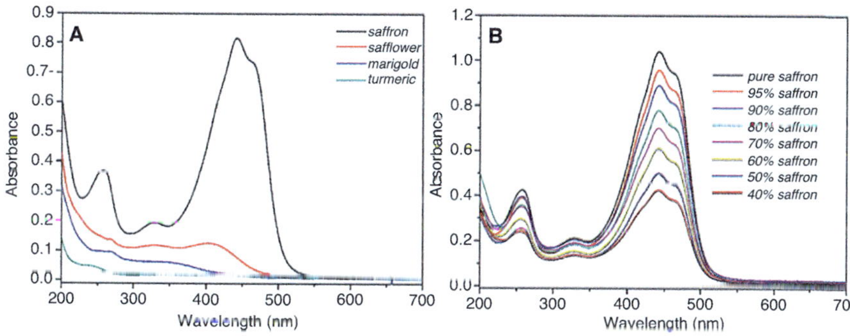

Fig. 4.5. UV-Vis spectra of saffron, safflower, marigold and turmeric extracts prepared using the reference method (A) and the reduction of absorbance with increasing adulteration ratios in the range of 5–60% (B) (Varliklioz *et al.*, 2017). Figure used with permission from Springer.

usefulness for intercomparison within trade, and then requires samples of interest to be analysed more specifically with targeted analysis techniques such as HPLC/GC-MS or conceptual fingerprinting with NMR analysis. The reasoning here aligns with the appendix supplement referenced in Salehi *et al.* (2022), wherein UV-Vis spectroscopy would be seen as useful but not alone sufficient for various authentic adulterations involving, for example, dyed materials, botanicals and mixing with degraded stigmas.

4.6.2 Thin-layer chromatography (TLC)

TLC represents a scientifically recognized method for saffron authentication, offering a useful alternative for pattern-level chemical trace evidence that goes beyond simple spectral indices. Although TLC might be considered 'simple', it should be remembered that authentic saffron methods demonstrate successful TLC not due to its supposed simpleness but because it allows a specific and repeatable two-dimensional chromatographic pattern that identifies unexpected bands and discrepancies in traditional bands intensities for natural saffron. Specifically, it should be recognized that image analysis and chemometric tools have significantly revived TLC as a spot-test technique from a qualitative method with semi-quantitative status toward authentication: an article by Valle García-Rodríguez *et al.* (2014) illustrates that TLC-image analysis with multivariate classification as a technique might successfully and quickly conclude on saffron authentication and detect, based on an overall lane pattern, an attempted adulteration type and amount.

When TLC papers are equipped with a UV scanner system, it is quite determined that both saffron major components are accompanied by artificial colourants. Despite the simplicity and low cost of thin-layer chromatography, its limited resolution can result in the overlapping of chemical spots, potentially leading to misidentification or overestimation of specific compounds. It becomes particularly appreciated within powdered saffron samples, wherein macroscopic observation appears ineffective, and mixtures are common within adulterations. As an alternative paradigm, TLC pattern analysis supported the observation that agronomic variables can influence the variability of composition, thus confirming once again the capability of TLC fingerprints to signal either the variability and authenticity of samples or adulterations, as it becomes fundamental within classification modelling, being necessary to distinguish 'natural variability of genuine saffron from fraud signatures' (Valle García-Rodríguez *et al.*, 2017b).

In validated methods, TLC might be viewed as a 'bridge technique', more selective on a chemical level than UV-Vis spectroscopy, but more efficient in terms of cost and higher throughput than LC/MS analysis, and useful for large numbers of samples for selecting which ones should undergo confirmatory HPLC/GC-MS analysis. This high-throughput chromatographic analysis would assist in reasonable subset sample selections for compound-resolved determination, as reflected as a strategy within laboratory practice related to controlling saffron (Fig. 4.6) (Sereshti *et al.*, 2018).

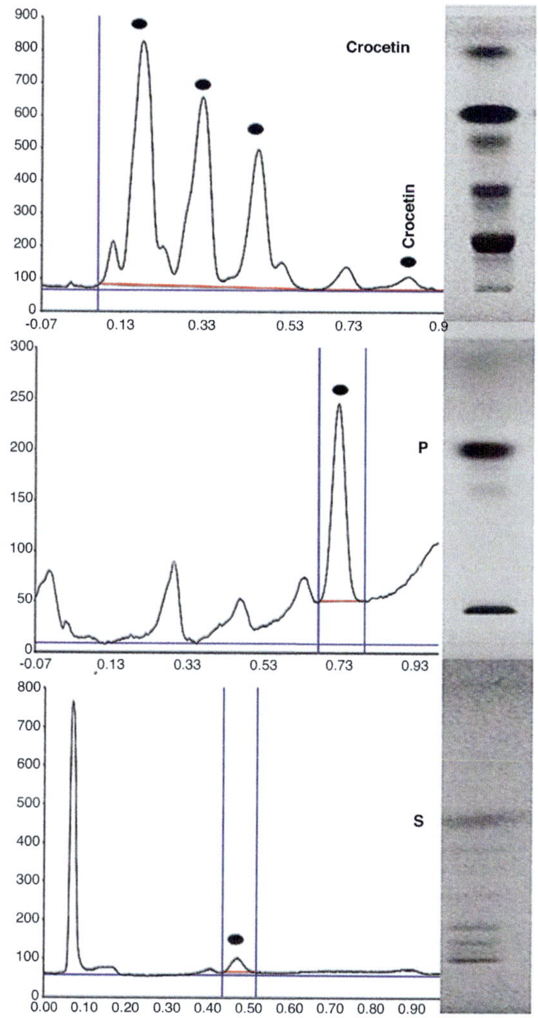

Fig. 4.6. HPTLC-based determination of total crocins (Crocin I, II, I0 and II0) and crocetin (top), picrocrocin (middle) and safranal (bottom) at 440 nm, 440 nm, 250 nm and 308 nm respectively (Kabiri *et al.*, 2017). Figure used with permission from Elsevier.

4.6.3 High-performance liquid chromatography (HPLC)

HPLC, often coupled with photodiode-array detection (PDA) or mass spectrometry (MS), allows the separation and quantification of individual crocins, picrocrocin and safranal. Reverse-phase HPLC is efficient in fingerprinting crocin derivatives. HPLC is generally considered the only backbone confirmatory method for the non-volatile constituents of saffron, due to the ability to separate and identify major compounds, which by definition are responsible for colour (the crocin series: multiple glycosylated crocetin esters) and bitterness (picrocrocin), further supporting more reliable evaluation of aroma-related constituents when used appropriately in multi-method schemes (Carmona *et al.*, 2006; Jarukas *et al.*, 2022). The chemical composition of the other *Crocus* species has been little studied and certainly offers extensive potential for future research. Figure 4.7 presents the chemical profiles of different saffron species. Interestingly, traces of crocins are typically detected in their methanol extracts, confirming the presence of an apocarotenoid metabolic pathway in these species. In the future, their genetic potential could be investigated.

A significant advantage of HPLC over the UV-Vis method is that the absorption values of the individual components separated by HPLC are recorded individually, making it much easier to detect adulteration of saffron. Studies published in ISI-indexed journals demonstrate that method validation in HPLC fully allows for the authentication of saffron's key markers, linearity, precision and robustness for major crocins and picrocrocin, enabling defensible quantification suitable for high-value trade, pharmacopoeial contexts and scientific comparisons across origins and harvest years (Carmona *et al.*, 2006).

HPLC is also of central importance for clarifying important quality aspects in saffron analysis: quality cannot generally be determined on the basis of a single crocin value alone. Saffron contains different crocins, which

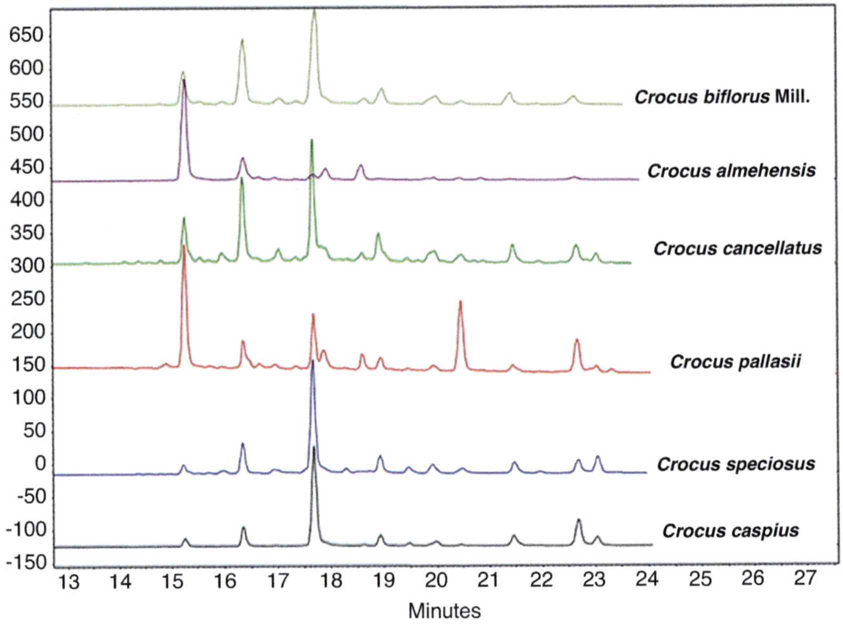

Fig. 4.7. HPLC profiles of various wild *Crocus* species recorded at 440 nm (Rezadoost, unpublished).

differ in terms of their respective glycosylation and isomerism. HPLC profiles therefore provide richer information, such as (i) the quantitative distribution of crocin types; (ii) the ratios between crocins and picrocrocin; and (iii) the occurrence of atypical profiles, which overall indicate adulteration with other plant species (e.g. pigments extracted from gardenias) (Fig. 4.8) (Kabiri *et al.*, 2017). For this reason, HPLC is often used as a reference when developing faster screening methods (spectroscopy, imaging, chemometric classification)

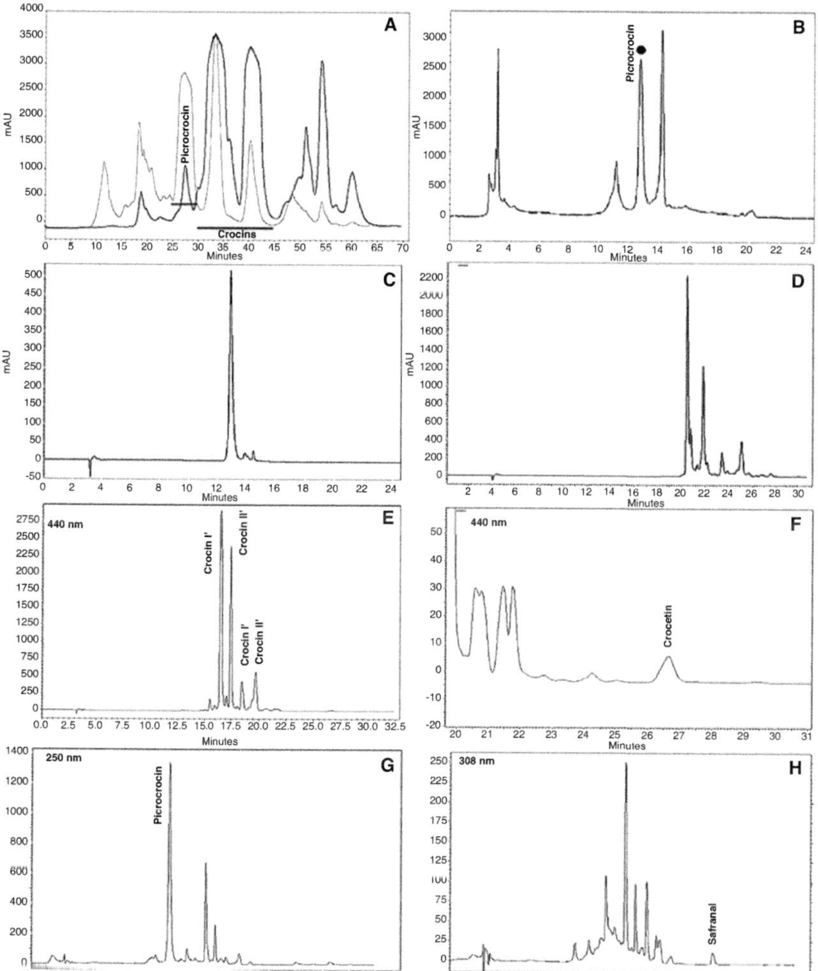

Fig. 4.8. Optimized preparative HPLC isolation of crocin mixtures (A) and picrocrocin (B) from an ethanolic extract of saffron. Analytical HPLC-DAD profiles of purified picrocrocin (C) and crocin mixtures (D) total crocins (crocins I, II, I′ and II′), detection at 440 nm (E), crocetin: 440 nm (F), picrocrocin: 250 nm (G) and safranal: 308 nm (H) (Kabiri *et al.*, 2017). Figure used with permission from Elsevier.

and when evaluating the reliability of standard UV-Vis indices, especially for aroma proxies (del Campo *et al.*, 2010; Jarukas *et al.*, 2022).

In addition to routine quality control, HPLC forms the basis for both profiling saffron metabolites and 'mapping chemical diversity', an aspect that is becoming increasingly important to explain authenticity, as there are natural variations in genuine saffron due to genotype, growing region and processing. Typical of this approach is the quantification of key metabolites across a large number of accessions and the use of comparisons of their profiles to explain deviations – an evidence base that enables more informed decisions about authenticity by defining the realistic chemical space of genuine material (Vahedi *et al.*, 2018). Finally, HPLC results are also particularly informative for the interface between product safety and authenticity: unusual peaks can be identified and then specifically tested for unauthorized colourants or impurities (e.g. by MS or NMR analyses). In this way, it is also possible to identify overstocked saffron samples, even if the external appearance of the medicinal product is unremarkable (Fig. 4.9).

4.6.4 Gas chromatography-mass spectrometry (GC-MS)

GC-MS is ideal for profiling volatile compounds, especially safranal and related terpenes. GC-MS headspace techniques further improve sensitivity and reduce matrix interferences. GC-MS is an important validated method for better understanding the volatile components of saffron, which are crucial for determining aroma quality influenced by various factors such as dehydration, overlaying, adulterants (e.g. mixing existing older stocks or certain processing methods to enhance the aroma impression). Safranal is generally the compound most reliably associated with the aroma of saffron, and GC methods have been proposed as superior for classification because they can directly measure the volatile components relevant to aroma. Bononi and colleagues (2015) compared GC data on safranal with the ISO-absorption method and suggested GC as the method of choice in this context.

Apart from single component analysis, the pattern of volatile fingerprints is also playing an increasingly important role. It has been shown that the profile of volatile compounds contributes to the fingerprint of aroma compounds, which may contain information about freshness, origin and adulteration. Farag and colleagues (2020) have shown that indices for freshness, origin and adulteration with related plants can be determined based on GC-MS aroma patterns using chemometric analyses, revealing a new trend in which pattern analysis significantly improves authenticity even in complex forms of adulteration. This is particularly evident in the studies by Maggi *et al.* (2010), who documented changes in the volatile profiles of different saffron samples depending on the duration of storage, thereby demonstrating why factors identified at the time of harvest should always be taken into account when testing aroma authenticity and why GC-MS analysis is fundamentally capable of distinguishing between 'quality reduction' and 'fraud' in the event of changes in the aroma profile.

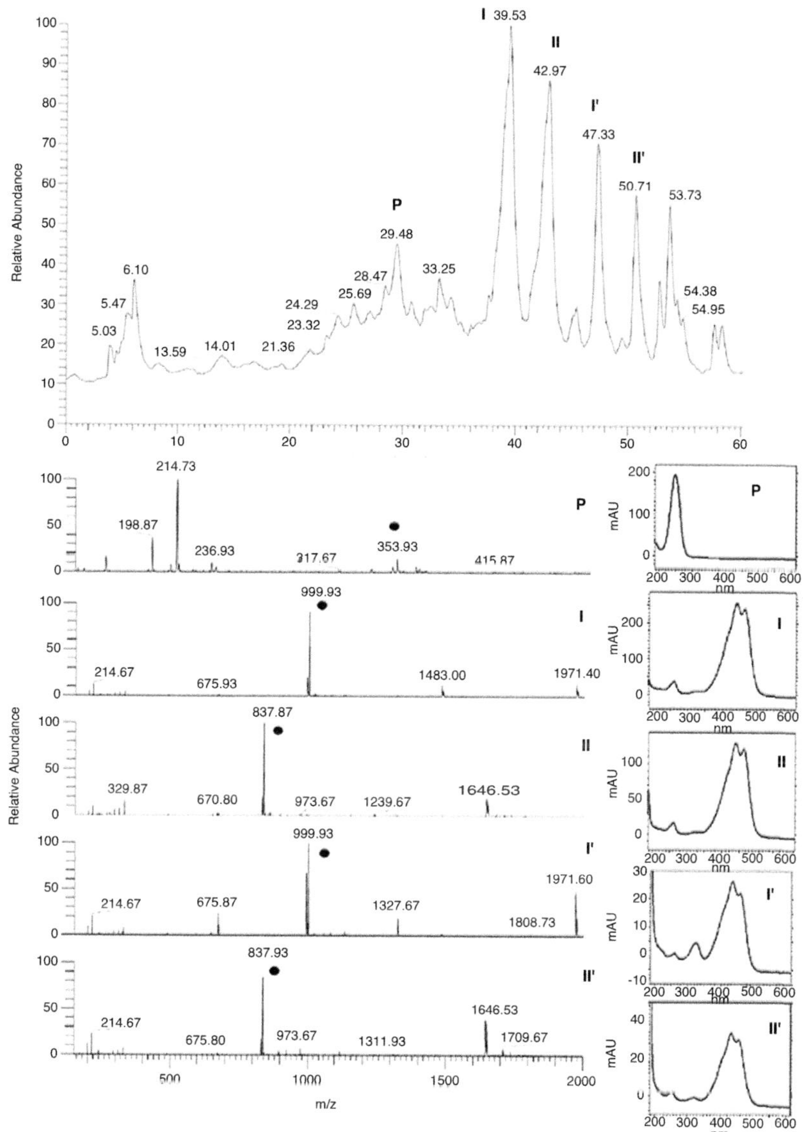

Fig. 4.9. HPLC-DAD-MS profiling of saffron extract. According to both mass spectra and UV-Vis spectra the presence of picrocrocin (P), crocin I (I), crocin II (II), crocin I′ and crocin II′ are confirmed (Kabiri *et al.*, 2017). Figure used with permission from Elsevier.

GC-MS is also becoming increasingly important as a method for authenticating saffron powder samples, as authentication at the morphological level is not possible here (Filatova *et al.*, 2024). In addition, GC-MS also plays a complementary role to HPLC in validated control systems, so that volatile substances can be taken into account as markers alongside non-volatile substances such as various crocins and picrocrocin for aroma (Fig. 4.10).

4.6.5 Nuclear magnetic resonance (NMR)

NMR spectroscopy is generally applied to elucidate the structure of new or modified saffron constituents. Nevertheless, it can also be used to support purity assessments and degradation studies. Today, NMR is widely regarded as one of the most efficient validated methods for authenticating saffron, as it provides an extremely reliable and detailed metabolic fingerprint. In saffron testing, NMR is typically used in two complementary ways: as a targeted confirmatory analysis for specific prohibited adulterants and as a non-targeted fingerprint analysis based on anomaly detection. A paradigmatic example of targeted analysis is the identification and quantification of Sudan dyes as a ^{1}H-NMR method for contaminated saffron samples, which is crucial evidence that NMR can effectively address the security issue related to fraud. Authenticity classification, based on learning the normal metabolic fingerprint of genuine saffron and identifying samples that deviate from it due to botanical adulterants, fillers and atypical processing, is enabled in non-targeted mode by NMR and multivariate statistical analysis (metabolomics). A strategy combining metabolomics and spectrometry, including H-NMR spectroscopy,

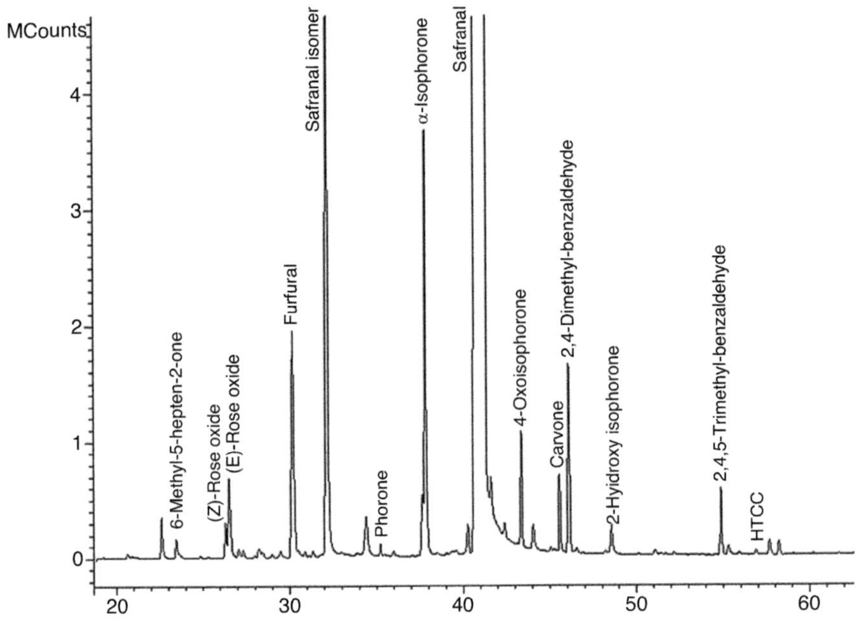

Fig. 4.10. SPME-GC-MS chromatogram of a Sicilian saffron sample (Condurso *et al.*, 2016). Figure used with permission from Taylor & Francis.

for traceability and the resolution of issues related to authentication and authenticity classification, particularly with regard to the ageing/quality classification of commercial samples of undetermined origin, has been proposed by Consonni and colleagues (2016) (Fig. 4.11).

Research falls perfectly within this proof paradigm because it shows that ^{1}H NMR analysis with data mining is capable of tracing authentication patterns and distinguishing genuine samples from fraudulent ones in saffron, thus again confirming NMR as an authentic verification method with low ambiguity on GC and screening tests (Dowlatabadi *et al.*, 2017). A particular advantage of NMR fingerprinting methods is that they offer the possibility of identifying different types of anomalies simultaneously (unusual aromatic rings, sugar patterns, additives). NMR spectroscopy with modern multi-stage methods is therefore now in demand in court as an extremely reliable technique, while UV-Vis and TLC are used more for preliminary screening and HPLC and GC-MS analyses for specific applications (Salehi *et al.*, 2022).

4.7 Authentication and Detection of Adulterants

Since saffron is expensive and requires labour-intensive harvesting, it is frequently adulterated with cheaper materials such as artificial colourings, non-food fibres derived from plants and inferior saffron. However, due to the harmful

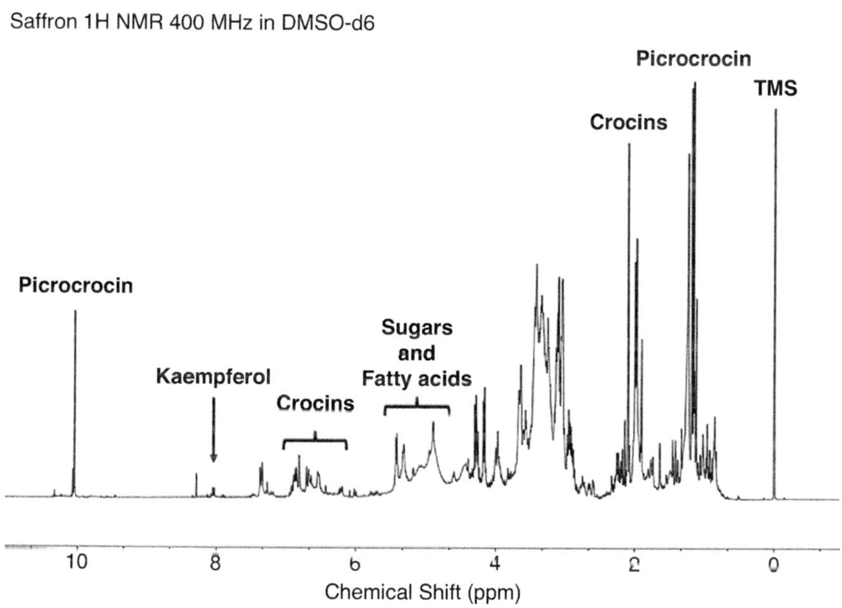

Fig. 4.11. 1H NMR spectrum of an authentic saffron sample (Angeli *et al.*, 2026). Figure used with permission from Elsevier.

effects on health and the deception of customers, particular attention must be paid to possible adulteration, as the adverse effects of these additives may lead to life-threatening allergies, intestinal disorders and irreversible damage to health.

At the same time, the use of inferior saffron instead of expensive saffron can trigger allergies in certain individuals. Apart from health risks, the adulteration of saffron also undermines historical cultural traditions that are still practised in Iran, India and Spain. According to the traditions and rituals practised in these regions, the use of saffron is indispensable. In addition, saffron acts as an antidepressant and antioxidant. However, adulteration largely negates its medicinal benefits and can even lead to illness.

To solve this problem, various techniques have been introduced to authenticate saffron and detect adulteration. Sensory examination techniques, which test the colour, aroma and taste of the drug, for example, are generally not sufficient to clearly identify adulteration. More specific and reliable techniques include chromatographic analysis methods, such as high-performance liquid chromatography and thin-layer chromatography. These technologies detect and separate the specific chemicals found in pure saffron. Non-destructive spectroscopic methods such as visible and infrared spectroscopy and Raman spectroscopy also offer efficient techniques for verifying the basic components of saffron. Moreover, DNA barcoding based on polymerase chain reaction (PCR) allows detection and identification of specific DNA regions associated with saffron. Additionally, microscopic examination enables detection and identification of extraneous matters added with the threads of saffron.

4.7.1 Hyperspectral imaging (HSI)

Hyperspectral imaging (HSI) and its technique have emerged as among the most attractive active front-line technologies for authenticating saffron because it integrates spatial information (the location within a sample) and spectral information (the composition of materials), thus enabling authentication on not only an average but also at local and dispersed levels.

In the case of saffron, adulteration can take the form of dyed threads, the addition of flower components such as pistils, and local ingredients from other plant species such as turmeric, safflower and marigold (Kiani *et al.*, 2023).

Validation tests have demonstrated various successful analyses based on near-infrared HSI and appropriate statistical models for the precise identification and quantification of adulterations in saffron. For example, researchers from Malawi have demonstrated successful NIR-HSI multispectral imaging and modelling to quantify and determine the extent of adulteration of *Crocus sativus* in varying proportions on saffron threads (Malavi *et al.*, 2024).

Moreover, general works on HSI images within the Vis-NIR range have shown that various plant-based adulterators, including safflower, style, calendula, rubia and turmeric, can be detected and classification can be performed for authentic and adulterated samples using classification with supervision, thus emphasizing the generalizability principle associated with the method (Hashemi-Nasab and Parastar, 2022).

Since the occurrence and distribution of crocins in saffron samples vary and are known to depend on the drying and processing background, spectral imaging can be used to make the individual carotenoids spatially visible. In this context, HSI can be used not only to determine whether a sample is genuine saffron, but also whether the saffron sample under examination is consistent and uniform in terms of its important substances. HSI is therefore an extremely valuable dual-purpose tool that plays a central role in modern authentication analysis. Related research in the field of spectral imaging for saffron samples, which uses multispectral Vis-NIR imaging as opposed to HSI imaging, has shown similar success. Alighaleh *et al.* (2023) demonstrated that spectral imaging features evaluated using machine learning classifiers could effectively distinguish genuine saffron samples from various types of counterfeits, including dyed fibres and samples consisting of safflower materials. Research in this area focuses particularly on operational performance, with HSI spectral data features typically combined with other imaging techniques to provide reliable multivariate analysis as a 'decision aid' for authentication purposes (Song *et al.*, 2024) (Fig. 4.12).

Nevertheless, there are several scientific factors in the existing literature on saffron that are considered decisive within an HSI in determining whether it is of authenticatable quality.

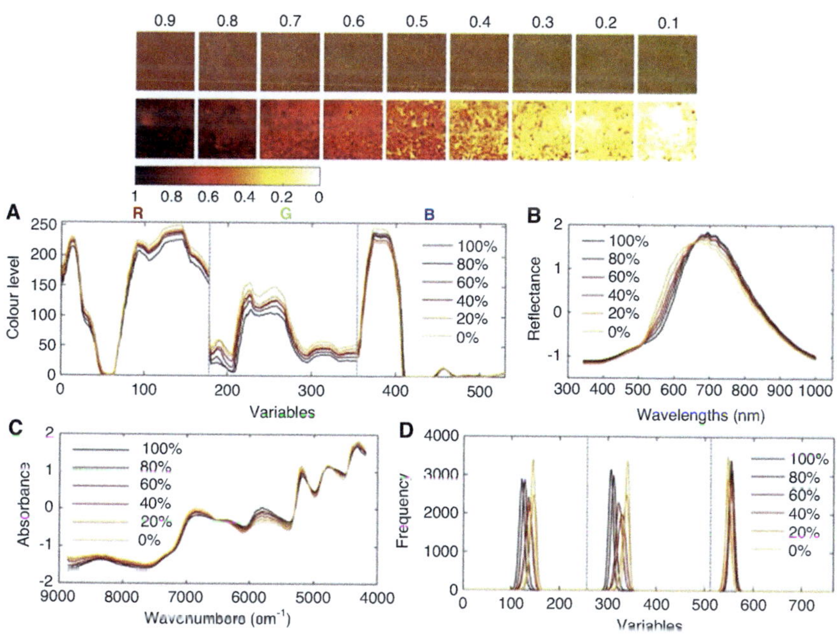

Fig. 4.12. Rapid and low-cost detection of saffron (*Crocus sativus*) adulteration using smartphone videos and spectral data fusion strategy (smartphone (a), HSI (b), NIR spectroscopy (c) and Computer vision system (d)) (Song *et al.*, 2024). Figure used with permission from Elsevier.

First, there are natural variability factors within saffron itself so that a narrowly defined HSI may classify atypical but genuine samples as authentic. Therefore, there is need for representation and verification outside this framework (Hashemi-Nasab and Parastar, 2022; Malavi *et al.*, 2024). Second, a low degree of adulteration may be difficult to identify, especially if the adulteration particles are unevenly distributed. Thus, testing needs to be done within a more realistic distribution and not solely on laboratory mixtures (Malavi *et al.*, 2024).

Third, HSI does not replace reference methods but serves as a screening procedure with classifications that are linked to traditional techniques such as HPLC/GC-MS and NMR. A large number of research studies are addressing precisely these considerations and using traditional, chemistry-based quantification methods to validate predictions about what promotes discrimination based on chemical differences (Alighaleh *et al.*, 2023).

Finally, interpretability is becoming increasingly important, as beyond accurate classification, the most valuable HSI tools will be those that can explain which spectral bands a classification result is based on with respect to crocin or water/moisture and dyes/adulterants.

Overall, HSI authentication tools have proven to be highly credible, as HSI takes a unique approach to identifying adulteration as a spatial-chemical problem rather than an average spectral problem, and as it is a fast and non-destructive method for combating complex fraud techniques (Hashemi-Nasab and Parastar, 2022; Alighaleh *et al.*, 2023; Malavi *et al.*, 2024; Song *et al.*, 2024).

4.7.2 DNA barcoding

As already mentioned above, adulteration of saffron cannot usually be detected by visual inspection; identification is practically impossible, especially in powdered saffron samples. In such cases, DNA barcoding using standardized genetic loci can be used to confirm that a sample actually originates from *Crocus sativus* (Bhooma *et al.*, 2023). Validated studies show that DNA barcoding can be used to successfully distinguish saffron from common adulterants and identify unauthorized additives, even when chemical tests are not able to detect these adulterants. A representative study has shown how DNA barcoding can be applied to distinguish *C. sativus* from various adulterants (Huang *et al.*, 2015). Barcoding targets particularly those genome regions that show sufficient differences between individual plant species and that can be amplified with universal or semi-universal primers. In the case of saffron, the internal transcribed spacer (ITS)-derived markers (a genomic sequence commonly utilized for phylogenetic analyses and species identification) are particularly relevant for plant authentication, as they have greater discriminatory power between close relatives and common botanical adulterants (Huang *et al.*, 2015). However, it should be mentioned here that DNA barcoding does not simply represent a yes/no identification approach. Villa *et al.* (2017) provide a clear example of this from the field of food chemistry. They developed and validated species-specific PCR and real-time PCR assays targeting the ITS region to detect and quantify adulteration of saffron with safflower (*Carthamus tinctorius*), as one of the most common saffron adulterants worldwide, including the ability to detect even very low levels associated with significant adulteration or cross-contamination (Villa *et al.*, 2017).

Khilare *et al.* (2019) proposed a three-step strategy combining ISO-standard quality assessment with microscopy and DNA barcoding, concluding that no single method adequately covers both quality and authenticity and that barcoding plays a crucial role overall by confirming the individual botanical origin.

However, DNA barcoding also has some scientific limitations that should be taken into account when applying it. Since it is known that drying and storage partially degrade saffron DNA, and pigments and other components can inhibit amplification, it is not always possible to obtain clear results when applying DNA barcoding.

4.7.3 Chemometric approaches

As already mentioned several times before, it is generally difficult to develop efficient quality control methods with thresholds for multiple variables simultaneously for saffron. Therefore, in such complex cases, today various statistical pattern recognition methods such as principal component analysis (PCA) are commonly used to accurately describe natural clusters in terms of valuable constituents based on origin, age and processed samples. These methods are particularly efficient for identifying outliers based on spectral and chromatographic variability. Consonni *et al.* (2016) describe chemometric tools such as PCA and discriminant analysis for saffron based on spectral data classified according to variables such as age and quality status. This typically allows the definition of authentic thresholds based on a pattern recognition approach (Consonni *et al.*, 2016). Similarly, NMR-based chemometric research studies have shown that multivariate analysis can identify patterns of adulteration within genuine samples of saffron based on understanding the 'metabolic range' of authentic samples and identifying discrepancies that correspond with adulteration with foreign materials (Dowlatabadi *et al.*, 2017).

In the field of rapid authenticity testing, chemometric methods are indispensable in hyperspectral or IR imaging systems, as the spectra provide a wealth of information that cannot be interpreted without the use of powerful evaluation algorithms. Also, very recent work on HSI (hyperspectral imaging) with a focus on saffron clearly illustrates that classification and quantification performance can be greatly affected by the kind of modelling approaches and variable reduction methods and suggests again that it is absolutely vital to have what has been termed a 'chemometric layer' as it contains the central analytic functionality (Malavi *et al.*, 2024). A similar observation can be made regarding IR/FTIR methods for distinguishing saffron authenticity. Solutions based on chemometric FTIR methods were described by Foschi *et al.* (2023) as discrimination classes for saffron adulteration, including older saffron threads and adulterants using supervised learning algorithms.

A further critical contribution from chemometric tools regards traceability modelling, wherein authentication aims not only at detecting adulteration but also at assigning samples according to geographical origin and production method. Moreover, emergent research on alternative analytical fingerprints (such as ion mobility spectrometry [IMS] or GC-based headspace fingerprinting methods) continues to be founded on classification tools; for example, recent verification research has shown the rapid possibility of geographical authentication methods on separating saffron origins within and among different regions (Fig. 4.13) (Shamshiri *et al.*, 2024).

While the type of applied statistical technique can affect the results, saffron analysis based on chemometrics requires special scientific precision. Overfitting is a recognized problem that might occur if classification models learn data artefacts instead of valuable chemical parameters due to inadequate cross-validation. Thus, serious research papers stress the importance of external validation sets and sound cross-validation designs and indicators (for instance, loading plot visualization and variable importance as a method of testing if learned regions appear meaningful from a chemistry viewpoint). In chemometric analysis, it is also necessary to define suitable training sets that reflect the individual differences between different saffron samples. Research into saffron authentication is increasingly relying on data fusion logic and reasoning tools, as these strengthen confidence in the presence of ambiguous adulterants (Alighaleh *et al.*, 2023).

In summary, chemometrics is a powerful scalable authentication method that makes it possible to check saffron samples offered in retail shops and online marketplaces. It helps to interpret large data sets as identification and

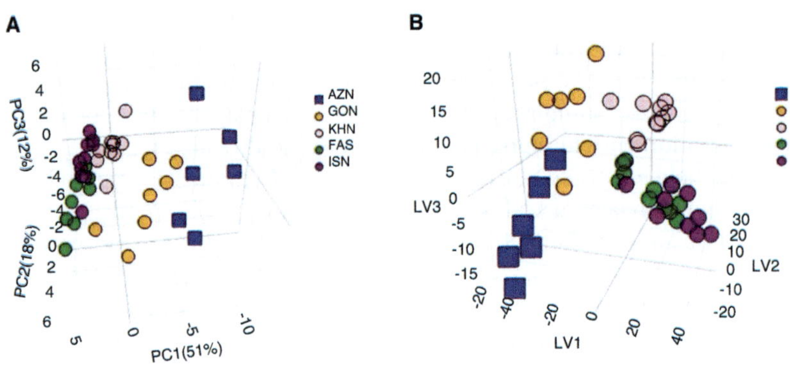

Fig. 4.13. Discrimination of saffron stigmas from different provinces, Azerbaijan (AZN), Golestan (GON), Khorasan (KHN), Fars (FAS) and Isfahan (ISN) of Iran. (A) PCA scores plot (PC1 vs. PC2 vs. PC3) in the analysis of the IMS spectra of saffron stigmas; (B) PLS-DA scores plot (LV1 vs. LV2 vs. LV2) on the selected (VIP > 1) data set. PCA: principal component analysis; IMS: ion mobility spectrometry; PLS-DA: partial least squares discriminant analysis; LV: latent variables; VIP: variable importance in projection (Shamshiri *et al.*, 2024). Figure used with permission from Elsevier.

quantification tools, thereby helping to avoid false positives and false negatives (Consonni *et al.*, 2016; Dowlatabadi *et al.*, 2017; Foschi *et al.*, 2023; Malavi *et al.*, 2024; Shamshiri *et al.*, 2024).

4.8 Phytochemical Differentiation by Geographical Origin

Differentiation based on geographical origin is thus scientifically plausible for saffron, given that its defining chemistry is influenced both by pre-harvest environment and post-harvest handling, and these factors vary systematically with region. Altitude, temperature amplitude, soil mineral profile, irrigation regime and harvest timing all may affect carotenoid/apocarotenoid metabolism in the stigma, while drying and storage practices greatly impact the formation and preservation of aroma-related constituents – most notably, safranal – and the stability of crocins. Accordingly, authentic saffron of different producing regions will frequently exhibit characteristic patterns in the composition and abundance of crocins, in the level of picrocrocin and in volatile fingerprints, even though all genuine saffron originates from the same species. One of the most comprehensive early chemical studies on variability between different places of origin is the study by Caballero-Ortega *et al.* (2007), in which the main components of saffron were quantified in certified samples from several countries and significant differences between places of origin were found in terms of the concentration profiles of important metabolites. A second line of evidence comes from integrated, multi-parameter origin studies, which combine chemistry with broader authenticity markers. Maggi *et al.* (2011) performed a preliminary investigation across authentic saffron samples from Greece, Iran, Italy and Spain using stable isotope composition together with chemical characterization by UV-Vis, HPLC and GC, illustrating how origin differentiation can be strengthened when compositional chemistry is interpreted alongside environmental signatures captured by isotopic patterns. This is supplemented by a study of volatile substances, which confirms the idea that signatures of origin are not limited to crocins: Masi *et al.* (2016) characterized saffron from Italy and Iran using aroma analysis in conjunction with HPLC profiling, supporting the idea that origin differentiation can be achieved both through bioactive compound profiles and aroma fingerprints, each reflecting different aspects of cultivation and processing traditions.

From a chemometric perspective, geographical differentiation becomes most robust if multivariate models learn entire fingerprints rather than single markers. A classic example is the mid-infrared spectroscopy study by Anastasaki *et al.* (2010), who, through multivariate analysis of IR spectra, were able to discriminate, with high correct classification rates, saffron samples from Greece, Iran, Italy and Spain, thus showing that origin information is embedded in broad compositional signatures detectable by spectral fingerprints.

Following studies have continued this direction by optimizing headspace/volatile profiling together with multivariate marker selection to capture geographical volatile markers and by developing integrated workflows, which combine multiple analytical domains to improve the confidence in discrimination, especially when different regions share similar climatic conditions or their post-harvest practices converge (Fig. 4.14) (Lu *et al.*, 2024).

The crucial point of interpretation for an introduction is therefore that the differences between 'Iranian vs. Spanish vs. Greek vs. Kashmiri' are not absolute chemical values, but statistical trends across populations of authentic samples: each region has an internal range, and overlaps can occur due to common varieties, similar drying methods or mixing in trade. This means that scientifically credible traceability of origin increasingly relies on well-curated authentic reference sets, multi-marker fingerprints and chemometric validation against confounding factors such as storage age or moisture differences. This explains why modern origin studies generally incorporate chemical profiles into pattern recognition rather than relying on a single threshold for crocin or safranal. Another development is the inclusion of multi-elemental and isotopic data, sometimes combined with chemical profiling, that provides signals that are harder to falsify as they reflect the individual geology and agricultural environment. These recent studies, which combine multi-element and isotope analysis with chemometrics, explicitly aim to distinguish between different geographical origins and illustrate the continuous expansion of tools for tracing the origin of saffron (Terro *et al.*, 2025). HSI will also be included

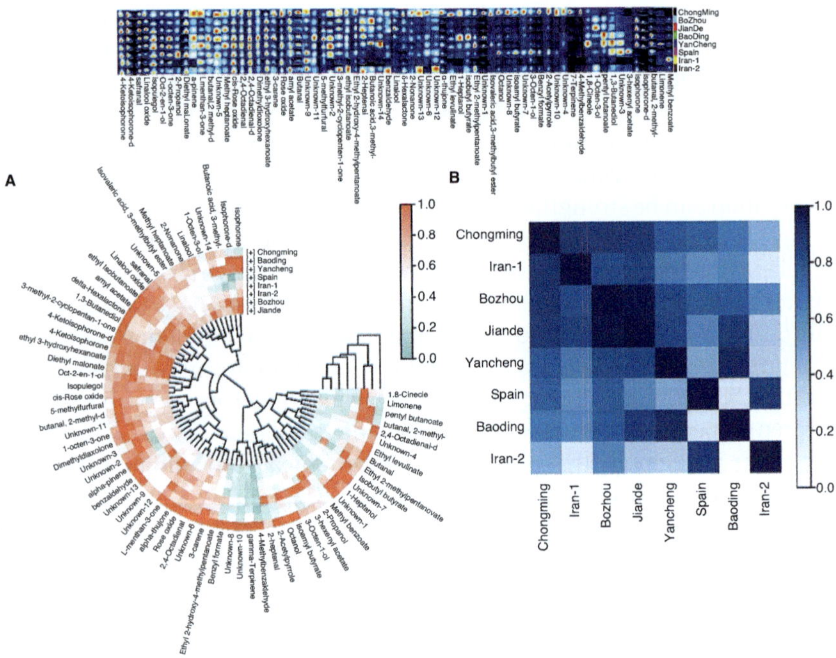

Fig. 4.14. Top: The fingerprint of volatile organic compounds (VOCs) detected in 8 different origins. Bottom: (A) Hierarchical cluster analysis of VOCs in saffron samples from 8 different origins. (B) Correlation analysis of 8 different saffron origins (Lu *et al.*, 2024). Figure used with permission from Elsevier.

in the geographical discussion, not only as a detector of adulteration, but also as a distinguishing feature for quality and origin. Portable spectral imaging systems (e.g. Vis-NIR spectral instruments) have been investigated for origin differentiation and quality determination, reflecting a trend towards non-destructive tools that can support traceability early in the supply chain, for example in procurement, export control or classification at the cooperative level (Caballero-Ortega *et al.*, 2007; Anastasaki *et al.*, 2010; Maggi *et al.*, 2011; Masi *et al.*, 2016; Lu *et al.*, 2024; Terro *et al.*, 2025).

4.9 Conclusion

Saffron is a chemically and biologically unique spice whose value depends on a highly integrated value chain ranging from specialization in stigmas and biosynthesis to post-production conversion and multi-layered analytical measurements. The available literature shows that the value-adding metabolites of saffron, namely crocins, picrocrocin and safranal, are not separate measures of product quality, but are linked via an apocarotenoid biosynthesis pathway, which itself derives from a universal carotenoid metabolism and proceeds enzymatically via CCD2, followed by an oxidation and glycosylation reaction and finally sequestered via transport functions. From a methodological perspective, numerous studies agree on a useful consensus: no single method can effectively combat fraud in the saffron markets, so multi-stage and orthogonal methods are most successful. UV-Vis and TLC/HPTLC analyses enable rapid and cost-effective preliminary fraud detection and pattern recognition, but must be improved in certain cases. Nevertheless, HPLC analysis (in conjunction with DAD/MS analysis) remains the backbone for legitimate quantification and the establishment of a reference space based on authentic variability and procedures. GC-MS analysis, on the other hand, is the most suitable technique for verifying volatile ingredients and relates the authentic composition of aroma compounds to drying and preservation, as well as the detection of fraud patterns using pigment-based methods. NMR, on the other hand, is an exceptional additional tool for authenticity and safety control, as it enables holistic metabolomic analysis and targeted analysis of fraudulent additives based on synthetic dyes.

Finally, recent studies show that authentication methods for saffron are increasingly focusing on a data-driven fingerprint approach, in which chemometric tools validate complex chromatographic and spectrographic data with models for classification and quantification. At the forefront of these developments is hyperspectral imaging, which, due to its focus on authenticity as a spatial-chemical problem, clearly represents an attractive and non destructive analysis option, enabling the detection of heterogeneous adulterations at batch level, while separate but complementary authentication methods focus on DNA barcoding/PCR as a solution for processed samples where morphology and chemistry are manipulable as integrity checks.

5 Industrial Applications: Use in Food, Cosmetics and Pharmaceutical Products

Hassan Rezadoost and Hartwig Schulz

5.1 Introduction

In order to explain the various uses of saffron in different areas, it is essential to know in detail about biochemical and functional properties of this drug. Saffron has different specifications in its three main areas of application, which form the basis for the use of saffron in the food, pharmaceutical and cosmetics industries. Despite many years of intensive research, the exact molecular substances responsible for the various bioactive properties of saffron are still largely unknown. The most important bioactive compounds in saffron, namely crocin, picrocrocin and safranal, have been studied for their individual antioxidant, anti-inflammatory, neuroprotective and anti-cancer properties. Synergistic effects have also been observed in this context, particularly between crocin and safranal.

This suggests that, in addition to the three molecules mentioned above, other substances occurring in saffron also contribute to its bioactivity, as was recently discovered (Si *et al.*, 2022). This chapter introduces the most important companies in the saffron industry and their commercial products to show how scientific findings have been successfully used in the development of functional foods, dietary supplements and cosmetic products. Through bridging scientific findings with industrial practices, it has been ensured that this chapter will stress the increasing relevance of saffron, which has evolved from a common kitchen spice to a bioactive substance with enormous commercial and medicinal potential. Table 5.1 shows the wide range of biological activities attributed to saffron and its various constituents. These studies range from *in vitro* experiments, *in vivo* studies using animal models, to observations in humans. It is also important to note that some of these effects are mentioned in traditional medical texts, which have long been a part of understanding saffron's use in alternative medicine.

Table 5.1. Range of biological activities from *in vitro* experiments, *in vivo* studies using animal models, to human-based observations.

Pharmacological activity	Model/System	Mechanism of action	Key findings/Effects	Reference
Learning and memory enhancement	Animal and cellular (hippocampal neurons, rats)	Modulation of NMDA receptors; ↑ antioxidant enzymes (SOD, GSH)	Prevented scopolamine- and stress-induced learning deficits; improved spatial memory	Hosseinzadeh and Ziaee, 2006; Pitsikas *et al.*, 2007; Ghadrdoost *et al.*, 2011
Anti-Alzheimer's activity	*In vitro* and clinical (PC-12 cells, AD patients)	Inhibition of amyloid β aggregation; anti-apoptotic (↑Bcl-2/Bax); ↓acetylcholinesterase	Prevented neurodegeneration; saffron extracts improved cognitive function comparable to donepezil	Soeda *et al.*, 2001; Papandreou *et al.*, 2006; Akhondzadeh *et al.*, 2010b
Antidepressant and anxiolytic	Animal and clinical (mice, humans)	Inhibition of dopamine, norepinephrine and serotonin reuptake	Antidepressant efficacy comparable to imipramine and fluoxetine; mild anxiolytic effects	Hosseinzadeh *et al.*, 2004; Pitsikas *et al.*, 2008
Anti-addiction/morphine withdrawal	Mouse models	Interaction with the opioid system	Reduced withdrawal signs without motor suppression; decreased morphine CPP	Hosseinzadeh and Sadeghnia, 2007; Imenshahidi *et al.*, 2011
Cerebral ischaemia protection	*In vitro* and *in vivo*	Antioxidant and anti-apoptotic; inhibition of JNK and MMP-9 pathways	Reduced oxidative damage and ischaemia-induced neuronal death	Ochiai *et al.*, 2004; Zheng *et al.*, 2007; Hosseinzadeh *et al.*, 2012
Cardioprotective/anti-atherosclerotic	Quail, rats, endothelial cells	Antioxidant; inhibition of apoptosis; ↓Ox-LDL	Lowered serum TG, cholesterol and LDL; prevented atherosclerotic plaque formation	He *et al.*, 2005; Liu and Qian, 2005; Xu *et al.*, 2006

Continued

Table 5.1. Continued

Pharmacological activity	Model/System	Mechanism of action	Key findings/Effects	Reference
Anti-hyperlipidaemic/anti-hypertensive	Mice and rats	Inhibition of pancreatic lipase; modulation of lipid metabolism	Reduced TG and LDL; decreased arterial pressure in a dose-dependent manner	Lee *et al.*, 2005; Sheng *et al.*, 2006; Imenshahidi *et al.*, 2010
Sexual function enhancement	Rats	Possibly via dopamine and nitric oxide pathways	Increased sexual activity (mounting, erection frequency); reduced latency	Hosseinzadeh *et al.*, 2008; Shamsa *et al.*, 2009
Anti-tumour/anti-cancer	Mice, rats and cell lines	Induction of apoptosis; antioxidant and cytotoxic effects	Delayed tumour formation, reduced tumour size, extended lifespan; PEGylated crocin improved efficacy	Konoshima *et al.*, 1998; Rastgoo *et al.*, 2014
Hepatoprotective	Rats	Antioxidant; improved mitochondrial function	Prevented enzyme elevation and lipid peroxidation in toxin-induced liver injury	Lin and Wang, 1986; El-Maraghy *et al.*, 2009; El-Beshbishy *et al.*, 2012
Nephroprotective	Rats and mice	Antioxidant (↑SOD, GSH; ↓MDA)	Improved renal antioxidant status; reduced oxidative stress markers	Naghizadeh *et al.*, 2008; Chen *et al.*, 2010
Anti-inflammatory and analgesic	Mice and rats	COX-1 and COX-2 inhibition; modulation of immune response	Reduced oedema, inflammation and nociceptive pain; enhanced morphine analgesia	Xu *et al.*, 2009; Tamaddonfard and Hamzeh-Gooshchi, 2010

Ocular protection	Retinal cell and animal models	Antioxidant and anti-apoptotic	Improved retinal function in oxidative stress and macular degeneration models	Broadhead *et al.*, 2012; Bisti *et al.*, 2014
Antidotal/genoprotective	Rodent models	Antioxidant; DNA protection	Prevented diazinon- and acrylamide-induced oxidative and DNA damage	Hosseinzadeh and Ziaee, 2008; Hariri *et al.*, 2010
Antioxidant activity (general)	Multiple tissues and organs	Free radical scavenging; ↑ total thiols, SOD, GSH	Universal antioxidant protection across organs	Hosseinzadeh *et al.*, 2005; Hosseinzadeh and Noraei, 2009
Toxicity	Animal studies	—	Low toxicity; no significant organ damage even at high doses	Wang *et al.*, 1984; Hosseinzadeh *et al.*, 2010

The ↑ sign is scientific shorthand for 'increase' or 'upregulation'.

5.2 Major Pharmacological Activities of Crocin

Crocins, which belong to the apocarotenoids and consist mainly of crocin I, are bioactive compounds occurring in the stigmas of *Crocus sativus*. Although the total crocin content in the stigmas of *Crocus sativus* exceeds 40% of their dry weight, in absolute terms it is remarkably low compared to other plant-based raw materials such as *Gardenia jasminoides*, which contains barely more than 1% dry weight. Since separating and isolating individual crocin substances usually takes a lot of time, effort and money, mixtures of crocin homologues are often used in saffron products today.

In their article 'Comprehensive characterization of the constituents of *Crocus sativus* L. from different origins based on the combination of targeted and nontargeted strategies', Si *et al.* (2022) were recently able to classify 47 crocins into 8 different types, of which 32 compounds were newly identified, demonstrating the wide variety of different crocin derivatives in saffron. Numerous medicinal properties of crocin have been studied in various body systems. Some of the properties most frequently mentioned in the literature relate to the central nervous system. These include improved memory and learning ability, which has been demonstrated in both *in vitro* and *in vivo* studies. Furthermore, both *in vitro* studies and clinical data have demonstrated a positive effect in the treatment of Alzheimer's disease. Positive effects of saffron have also been observed in the treatment of morphine addiction and cerebral ischaemia. Finally, *in vivo* and clinical studies have shown that crocins have a positive effect on sexual function and even possess tumour-destroying properties. In fact, there has been considerable research done on the biological and pharmacological properties of two saffron components, safranal and crocin. Most of these studies fail to identify the types of apocarotenoids found in crocin. Also, it needs to be noted that total crocins in saffron stigma extract are around 40%. Saffron stigma extract and its predominant bioactive compounds, crocins, were found to possess various biological properties, including evidence from *in vitro* studies, *in vivo* experiments and human clinical trials.

Among the different therapeutic activities attributed to saffron, the effects of crocin on the central nervous system are the best documented and scientifically well validated. It acts by improving learning and memory, as indicated by *in vitro* and *in vivo* studies, and exerting positive effects on Alzheimer's disease, as demonstrated by *in vitro* studies and clinical data. Crocin was found to modulate morphine withdrawal, conditioned place preference and dependence in *in vivo* studies and to exert a neuroprotective effect against cerebral ischaemia in both *in vitro* and *in vivo* models. In addition to its effects on the central nervous system, it has also proven effective in a variety of therapeutic applications in various organ systems: for example, it has a positive effect on the cardiovascular system by acting against arteriosclerosis, hyperlipidaemia and high blood pressure *in vitro* and *in vivo*, and protects against heart muscle damage *in vivo*. Some studies on its effects on sexual function support the results of *in vivo* experiments and clinical trials. In addition, renoprotective (*in vivo*) and hepatoprotective (*in vivo*) effects have been reported. Its strong antioxidant effect has been demonstrated in *in vitro* and *in vivo* studies, and a therapeutic effect in eye diseases has also been demonstrated in *in vitro* and clinical studies. Other effects include anti-nociceptive and anti-inflammatory effects *in vitro* and *in vivo*, protection against ischaemic damage, and antidotes against toxins such as acrylamide and diazinon both *in vitro* and *in vivo*. Finally, its genoprotective effect has been observed in animals, suggesting that it has a variety of protective and healing effects. Although crocins, like other components of saffron, have a variety of beneficial biological effects, special attention must be paid to their safety and potential toxicity. Knowledge of the toxicological profile is therefore an important factor in ensuring the safe use of these compounds in experimental and clinical applications. Animal studies have shown that high amounts of extracts and isolated crocin may be responsible for certain toxic effects. A 14-day daily intraperitoneal administration of stigma and petal extracts to rats resulted in normochromic normocytic anaemia, while the petal extract also caused liver and lung toxicity (Karimi *et al.*, 2004). Hosseinzadeh and Nassiri-Asl (2013b) were able to demonstrate in acute studies on mice and rats that safranal is slightly toxic after intraperitoneal administration and practically non-toxic after oral administration. In subacute toxicity tests, safranal only altered certain haematological and biochemical parameters.

Clinical studies with saffron and crocin have shown that these substances are generally safe in healthy adults. One-week studies with daily doses of 200 or 400 mg of saffron did not cause any significant health impairments, although some haematological and biochemical values had changed slightly, but these values were within normal limits (Modaghegh *et al.*, 2008). In another 1-month study, only minor changes in haematological parameters were observed when 20 mg crocin tablets were administered to subjects, with no significant side effects (Mohammadpour *et al.*, 2013). Despite these results, larger, longer-term clinical studies are still needed to fully clarify the safety and health effects of saffron and its active ingredients. The carotenoid crocin is capable of scavenging free radicals, especially superoxide anions, thereby protecting cells from oxidative stress. Many reports describing the various effects of saffron on the central nervous system attribute these effects to the antioxidant properties of crocins (Abe and Saito, 2000).

Recent research reported the anti-cancer effects of safranal, which further demonstrated its impact on protein translation, mitochondrial dysfunction and DNA fragmentation. In this regard the phosphoproteomic and acetylomic analyses revealed that safranal modifies the signalling axis leading to the disruption of DNA replication and repair, translation and EGFR. DNA damage is induced, cell-cycle checkpoints are activated and then suppressed, which drives the cells to mitotic catastrophe and apoptosis. Further, safranal promotes cell death through the inhibition of DNA repair (by H4 hypoacetylation) and augmentation of proapoptotic gene expression (by H3 hyperacetylation) (Ashrafian *et al.*, 2022).

Abu Ali (Ibn) Sina, commonly known as Avicenna, was an Iranian physician and the most renowned and influential of the philosopher–scientists of Islam. Among many other works, he mentioned saffron in Book II of the *Canon of Medicine*. According to traditional medicine, saffron is warm and moderately drying. It is considered astringent, dissolving, promoting maturation in cases of swelling, and warming and strengthening for the internal organs. Saffron is traditionally used for cosmetic purposes, as it was believed that taking it orally improved the complexion. It was also used to dissolve swelling and treat certain inflammatory skin discolorations. In some cases it is beneficial, but excessive use can cause or exacerbate headaches, although it relieves warm inflammation of the ear. It also supports vision and is beneficial for day blindness and discoloration of certain parts of the eye. For stomach complaints, it was considered an emetic, appetite suppressant and a means of strengthening the stomach and liver. It was also used as a diuretic and aphrodisiac, as well as for gynaecological complaints such as uterine blockages and to ease labour pains (Hosseinzadeh and Nassiri-Asl, 2013a). The most important use of saffron was as a food colouring in all periods. Saffron has traditionally been used as a food colouring, a flavour and a fragrance in baked goods, desserts and other sweet and savoury dishes in Iranian and Indian cuisine. Moreover, it owes its place in many traditional dishes to this characteristic, including sholezard in Iranian cuisine, risotto alla milanese in Italian cuisine, paella in Spanish cuisine, Cornish saffron cake in British cuisine and lussekatter in Swedish cuisine (Sahakian, 2015).

However, saffron also has numerous other applications in the food, pharmaceutical and cosmetics industries, as well as being used as a textile dye. There are also saffron remedies for external use, for example for application to the skin. The application of saffron flowers to diseased areas of the body, as described by Bartholomaeus in *De proprietatibus rerum*, Book XVII, Chapter 41, is a topical application of saffron as a remedy. In terms of the pharmacological effects of saffron as reported in the literature, a total of 76 articles (from 630) and seven books (in English and Persian languages) were selected and compared with the plant monograph of Avicenna's *Canon of Medicine* (Hosseinzadeh and Nassiri-Asl, 2013a) (Fig. 5.1A).

There are numerous traditional medicines for bruises, where trauma causes blood to accumulate and clot under the skin without damaging the skin itself. Saffron is an important ingredient in these medicines. Recent preclinical studies in recent years have provided increasing evidence of the outstanding anti-cancer properties of the most important hydrated water-soluble carotenoids in saffron, known as crocins, in human skin cancer. Various *in vitro* studies using human squamous cell carcinoma skin cell lines, including A431 and SCL-1, have shown that crocin has a strong inhibitory effect on cell proliferation, induces apoptotic cell death and influences various carcinogenic signalling pathways, including JAK/STAT, Wnt/β-catenin and important miRNA-related pathways. These findings are further supported by studies in xenotransplanted models of human *in vivo* skin carcinomas,

Fig. 5.1. (A) Avicenna, the oldest scholar to use medical applications from saffron; (B) the traditional harvesting of saffron (Hosseinzadeh and Nassiri-Asl, 2013a); (C) 'Saffron Gatherers' fresco, discovered at Akrotiri on Thera/Santorini (Marinatos, 1985; Day, 2011a, b; Wikimedia Commons, 2026); (D) the use of saffron with rice, which is common in many countries around the world, including Iran and Italy. Figures (A) and (B) used with permission from Planta Medica.

which found that saffron crocin suppresses tumour development, alleviates inflammatory conditions and modulates oxidative stress. Saffron has also demonstrated enhanced anti-metastatic effects in melanoma, suggesting that saffron crocin has broader dermatological applications in cancer. New innovative approaches, including the use of nanoparticles as carriers, further improve the shelf life, bioavailability and potential for successful topical and/or systemic application of saffron crocin. Although such studies in human mammalian models are still limited, the results suggest that saffron crocin is a naturally effective molecule that has shown promising results in the prevention and treatment of skin cancer in humans.

5.3 Saffron Applications from Past to Present

5.3.1 Food applications

Archaeological and iconographic data suggest that saffron was already an established crop with recognized uses and value in the Aegean region during the Bronze Age. The famous 'Saffron Gatherers' fresco, discovered in Akrotiri on Thera/Santorini, depicts the harvesting of crocuses and is widely regarded as evidence that saffron was not an everyday food but a luxury product with ritual significance (Marinatos, 1985; Day, 2011a, b; Wikimedia Commons, 2026) (Fig. 5.1B, C). With regard to antiquity and the classical period, it should be noted that saffron went beyond its use as a symbol. It became a spice, a dye and a status symbol. Studies of ancient traditions mention that it was used in various civilizations as a spice that added colour and aroma to special occasions such as feasts with food and drink (Mzabri *et al.*, 2019). Historically, the use of saffron in cooking reached its peak in late antiquity and the Middle Ages, when trade routes were established from the Middle East and the Mediterranean to Europe. In the Middle Ages and early modern period, there was a town in England that later became known as Saffron Walden. Local historical texts often refer to it as an indication of the importance of saffron as a trade spice and more (Visit Saffron Walden, 2022).

In early modern times, saffron found its way into regional cuisine, from bouillabaisse-like fish stews and rice dishes to bread and sweet pastries to colourful and aromatic festive dishes (Fig. 5.1D). Contemporary culinary research emphasizes that saffron has a distinctive sensory 'signature pattern' – including colour, bitter compounds and aroma as core components – and therefore remains an indispensable ingredient despite the widespread availability of cheaper yellow colourings that cannot quite match it (Spence, 2023).

Due to its price, its use as a food product cannot be separated from control and authentication. The introduction of the international standard ISO 3632 makes it possible to objectively evaluate the colouring power of saffron using spectrophotometric values (International Organization for Standardization, 1993, 2010).

Current analytical methods for industry show how UV-Vis analysis works very well in testing for commercial classification based on current ISO standards (such as ISO 3632-1:2025) within the 'routine methods' (Heenatigala, 2025). At the same time, there are documented cases of adulteration in the current literature on food safety and a demand for reliable methods to detect it (Raina *et al.*, 2024).

5.3.2 Medical applications

Historically, the medical use of saffron coexisted with its esteemed culinary status. Historically, saffron was considered a 'high-value remedy' within traditional materia medica, associated with tonics and folk medicine and sometimes related to mood, energy and overall health and well-being, as with many aromatic herbs that were consumed as food and also have been used for medicinal purposes (Mzabri *et al.*, 2019; Lachguer *et al.*, 2025b). It is significant that it existed so long before modern times, as this explains why saffron was so intensively cultivated, preserved and traded, since saffron was more than just a spice for some cultures.

Moving on to modern medicine, the history of saffron changes from traditional usage to research trials. The most well-supported area with regards to the number of trials conducted would be mental health, with a focus on mild to moderate depression and anxiety. A more recent meta-analysis conducted on trials directly comparing saffron treatment with SSRI antidepressants showed that there was no significantly differing effect based on reduction of symptoms seen with either treatment, with fewer side effects seen with saffron, and a call for more comprehensive research on who would be best benefited by these therapies (Shafiee *et al.*, 2025). Clinical trials have also examined the use of saffron on specific populations. In a randomized double-blind study, saffron was compared with fluoxetine in the treatment of postpartum depression and showed equal efficacy within 6 weeks, although this was described as preliminary and insufficiently proven (Kashani *et al.*, 2017).

Another area of research concerns vision and retinal function, particularly age-related macular degeneration (AMD). A randomized, single-blind crossover study involving adults with mild/moderate AMD who received 20 mg of saffron daily for 3 months showed some benefits in terms of visual function compared to a placebo, with no differences in side effects (Broadhead *et al.*, 2019).

Although these findings do not qualify saffron as a single-agent treatment for AMD, they demonstrate a paradigm for modern medicine: that for a herbal supplement like saffron, there should be objective functional outcomes. Finally, the medicinal application of saffron today requires a considerable degree of standardization and authentication for two reasons. First, reference should be made to consistent doses and composition. Second, there have been instances of adulteration that have been discovered within the market associated with expensive botanicals. Current literature on food control and safety considers the problem of saffron adulteration to be a public health and consumer protection issue that applies equally to all medicines or food supplements where purity is important (Fig. 5.2) (Leone *et al.*, 2018).

5.3.3 Cosmetic applications

The cosmetic history of saffron begins in a place that is not surprising for many luxurious plant substances: rituals, beauty symbolism and personal care among the elite. Historical and ethnobotanical summaries report its use in skincare products – infusions, topical mixtures with fat or milk, traditional skincare products – in many cultures where the recurring themes are 'lightening', 'youthful appearance' and perfumed body care (Mzabri *et al.*, 2019). Even if the accuracy of certain anecdotes, for example about famous personalities, can hardly be historically verified, the pattern is unmistakable: the colour and scent of saffron made it a natural candidate for prestigious cosmetics. As perfumery evolved, saffron's role expanded from home preparations to fragrance structures. Recent reviews treat saffron as part of the vast aromatic plant repertoire destined for use in perfumery and cosmetology, and they also point

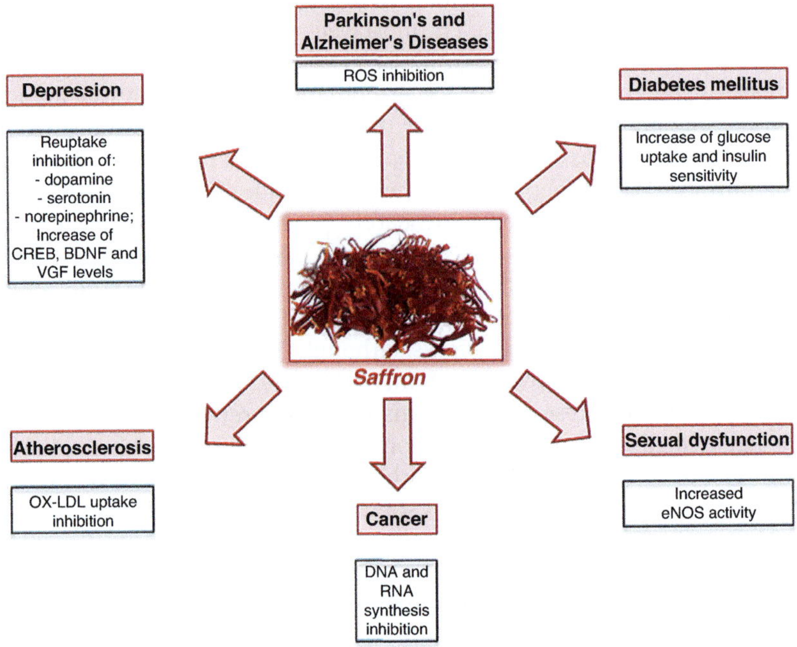

Fig. 5.2. Medical influences and applications of saffron (Leone *et al.*, 2018). Figure used with permission fom John Wiley & Sons.

out that saffron's appeal is sensorial as much as functional: it signals rarity, warmth and refinement in finished products (Mzabri *et al.*, 2019).

In modern cosmetics, so-called cosmeceuticals, saffron is positioned in three major ways:

i. *Experiential and branding ingredient:* a luxury feature in creams, serums, masks, soaps and perfumes, in which the cultural history is often used to reinforce the symbolic meaning of saffron.

ii. *Antioxidant/skin-comfort positioning:* contemporary product narratives often rely on antioxidant and skin-protective framing, borrowing from the wider scientific literature summarized in reviews.

iii. *Valorization of extracts as by-products:* a really 'now' development is the drive to utilize saffron flower by-products – primarily tepals/petals – as sources of phenolics and other bioactives for cosmetic formulations, in line with the circular economy. A review published in 2025 details their scale of by-products, which indicates the surprisingly rich profile of phenolics and pigments, pointing toward practical pathways into cosmetics and hygiene-related products (Lachguer *et al.*, 2025b).

It is not a theoretical by-product route either. The experimental work has indicated that the saffron tepals could be treated as a functional raw material whose antioxidant-related composition can be valorized with controlled growth conditions, such as manipulating light spectra, and would become phytochemically richer floral residues that might be channelled into streams for food, cosmetics or even pharmaceutical ingredients (Orlando *et al.*, 2022).

This is a significant modern shift: whereas in the past most of the value lay in the stigmata themselves, industrial innovation is increasingly seeking to find an answer to the question of how value can be extracted from the full bloom. Two practical constraints give a shape to cosmetics use today. First, there is standardization: the producers of cosmetics need reproducible extracts, for stability, colour and scent consistency, and this intersects with the same analytical quality frameworks used in the food trade – grading and authenticity testing (Bergomi *et al.*, 2022; Heenatigala, 2025). The second point is the risk of adulteration: since saffron is expensive, there are also incentives for substitution or mislabelling in the cosmetics supply chain, which is why strict testing is also necessary outside the food sector (Raina *et al.*, 2024). The significance of saffron cosmetics has shifted from traditional beauty rituals and fragrant status symbols to standardized, tested ingredients, with sustainability also playing an increasingly important role as petals/calyxes become a new raw material for modern formulations.

5.3.4 Industrial applications

As early as ancient times and the Bronze Age in the Aegean, the presence of the crocus symbol suggests that materials associated with saffron were so important that they appeared as artistic motifs, indicating an economy linked to luxury goods (Marinatos, 1985; Day, 2011b). Over the years, its value as a pigment made it a sought-after source of colour not only for food, but also for materials such as fabrics, ceremonial textiles and possibly inks or dyed media with yellow-orange tones that symbolized elegance and luxury (Mzabri *et al.*, 2019; Lachguer *et al.*, 2025a). Today, the industrial use of saffron can be summarized in four main areas:

i. *Classification and testing:* Due to the high costs and susceptibility to adulteration, there is a large 'saffron industry' associated with laboratories, standards, equipment, certification procedures and compliance with export regulations. The basis for the classification and testing of dried saffron was established by the ISO 3632 standard. Today's laboratory procedures are efficient and suitable for classification based on UV-Vis spectroscopy in compliance with current ISO documentation standards, as evidenced by reference standards published as applicable methods, including ISO 3632-1:2025 (Heenatigala, 2025). Rather, in recent years, authenticity checks have been supplemented by a combination of complementary methods such as Fourier Transform infrared spectroscopy, microscopy/SEM-EDX, gas chromatography and element profiling.

ii. *Detection and risk management of adulteration:* The industrial saffron market is constantly under fraud pressure, and it has been noticed that detection of adulteration is considered an important issue within modern food control literature (Raina *et al.*, 2024).

iii. *Natural dyes and materials:* Both dyes and pigment-rich by-products from saffron are often cited as natural sources for colouring. Although this would be too expensive as a source for general dyeing purposes, research and development in the field of sustainability is aimed at the use of specialized, high-quality natural dyes, particularly using methods based on the circular economy (Orlando *et al.*, 2022; Lachguer *et al.*, 2025a).

iv. *Utilization of by-products (circular economy):* A current trend in the industry is the expansion of the value chain for saffron beyond the stigmas. One report describes the large quantity of by-products (petals, stamens, etc.), and an overview describes possible uses for animal feed production, the extraction of phenols and pigments, and ingredients for various industries such as cosmetics and textiles (Lachguer *et al.*, 2025a). From an experimental point of view, there are reasons to favour this approach, as it has been shown that petals can be considered an active biological resource rather than waste material (Orlando *et al.*, 2022). Overall, the scope and use of saffron as an industrial product has shifted from traditional luxury goods and trade practices to a more contemporary sector that encompasses either qualitative analysis systems and methods for combating fraud or models for optimizing the entire flower, with the aim of making 'red gold' more sustainable (ISO, 1993; Bergomi *et al.*, 2022; Orlando *et al.*, 2022; Raina *et al.*, 2024; Heenatigala, 2025; Lachguer *et al.*, 2025a).

6 Global Saffron Market Dynamics, Socio-economic Aspects, International Quality Standards

Hassan Rezadoost and Hartwig Schulz

6.1 Why is Saffron Essential?

Saffron has been used for thousands of years, but recent studies suggest that this product still has a lot of untapped potential that has not yet been sufficiently researched. The diverse uses of saffron are due to the special properties of a group of compounds known as apocarotenoids, which include safranal, picrocrocin, crocins and crocetin. Industrially, the current use of saffron can be divided into four main areas: as a food ingredient, in herbal medicines, as a textile dye and in cosmetic products. These sectors correspond to established uses in the past and current uses (UNIDO, 2014). In addition to its traditional use as a spice in Middle Eastern, South Asian and Mediterranean cuisine, saffron is increasingly being used in new and value-added food systems, such as functional beverages, nutrafunctional dairy products, confectionery and nutraceutical formulations (Sanjuán-Lopez and Resano-Ezcaray, 2020). This is mainly due to changing consumer preferences for natural plant-based ingredients that not only improve the properties of food but also offer potential health benefits.

The commercial development of saffron as a medicinal product has not yet been fully completed, meaning that the current commercial formulations are still extremely limited (Kyriakoudi *et al.*, 2015b). In textile-related industries, saffron has been traditionally used as a natural dye with significant application in yielding vibrant yellow to reddish-gold shades. While saffron has lost momentum with the use of synthetic dyes, it still has niche application in specialized textile industries. The outstanding stability of saffron-coloured products contributes to saffron being used in the dye industry, albeit to a limited extent. Saffron has been used in perfumery and cosmetics for many years due to its characteristic scent and its associations with luxury and sophistication (Bathaie and Mousavi, 2010). Although saffron's volatile substances are used in certain perfumes, as well as traditional cosmetic formulations, their use on a mass industrial scale has been only partly developed. Overall, four areas (culinary arts, medicine, the textile industry and perfumery) constitute the most important established uses of saffron. Nevertheless, new scientific data and technological advances suggest that new uses are also emerging, particularly in the areas of functional foods, pharmaceuticals, cosmetics, aromatherapy, high-quality pigments and high-quality biomaterials. As a result, saffron is poised to gain prominence and has significant interdisciplinary and expansion potential. In summary, a number of key advantages have made saffron cultivation attractive not only to smallholders but also to commercial enterprises. This means that saffron farming does not necessarily require a lot of space, unlike other major agricultural products. In addition, saffron farming can even take place between orchards, and this ensures that any available space, including that between orchard trees, is utilized effectively. This particular advantage of saffron farming also ensures that it benefits small-scale farmers who lack agricultural space on their farms. In addition, saffron cultivation offers many employment opportunities, from harvesting to drying, sorting and packaging. These are all labour-intensive processes, which are therefore economically important for small economies. Here, too, the price–performance ratio is very good, so that even small farmers can benefit from selling saffron, even if only in small quantities, as this brings a reasonable economic advantage. In addition, saffron cultivation does not require any special equipment and can even be carried out without a large capital investment.

6.2 Saffron World Production

Iran, India (Kashmir), Spain (La Mancha), Greece (Kozani), Morocco and Afghanistan are among the major saffron-producing countries. With a share of about 90%, Iran is the biggest saffron-producing country with about 200–400 t/year production, estimated at 418 t/year on 121,338 ha (Tosan *et al.*, 2024). According to the United Nations Industrial Development Organization (UNIDO), in 2012, Iranian saffron production accounted for 90.7% of the total world production of saffron. With 78,804 ha under cultivation, Iran manufactures 261 t of saffron per year, reflecting an average yield of 3.3 kg/ha. The second largest producer, India-Kashmir, contributes 4.4% of global production and harvests 12.5 t from 3200 ha. Spain, historically a major saffron producer, now accounts for only 0.8% of global production, while Greece (Kozani) accounts for 2.5%. Individual producers, each accounting for less than 1%, include Italy (Sardinia), Afghanistan, Morocco, Azerbaijan and China. Total global production of saffron amounts to 288.4 t, grown on an area of 85,949 ha. In addition to official statistics, several recent market reports indicate that by 2030, Iran is projected to maintain an overwhelming majority share of global saffron production (roughly 80–90%), reflecting its continued leadership in output. Concurrently, the global saffron market is expected to expand substantially: independent forecasts suggest a market size in the order of US$ 0.9–1.2 billion by 2030, based on 2024 baselines and multi-annual growth rates (Jadhavar Business Intelligence, 2024).

Between 2016 and 2020, the global saffron trade was subject to significant fluctuations on both the export and import sides. Iran was consistently the leading exporter, reaching a peak of 351,127 kg in 2018, although its exports declined sharply in both 2019 and 2020, likely due to domestic consumption and trade restrictions (Table 6.1). Spain's export volume was moderate but stable, while Afghanistan was a rapidly emerging producer, with exports rising from 3869 kg in 2016 to 45,637 kg in 2020. Greece, France and Portugal were smaller exporters with declining volumes. On the import side, India showed an upward trend, reaching 23,019 kg in 2020, while the USA remained a stable market. Hong Kong was an important hub with highly fluctuating imports, while the United Arab Emirates recorded a sharp decline during this period. Overall, the data reflects the changes in the global saffron market, with Iran continuing to play the leading role.

6.3 Marketing Chains for Saffron: Structure and Policy Implications

Empirical studies on saffron value chains show that saffron markets are predominantly controlled by middlemen rather than by producers who sell their products directly or export them. In most saffron-growing regions, the majority of saffron produced is distributed through brokers and then through traders; however, a small percentage of sales are handled by subcontractors and other intermediaries. Current data show that 70.86% of saffron sales are handled by brokers, 16% by subcontractors and 13% by other intermediaries. This is also consistent with the findings of studies analysing the saffron value chain in the saffron-producing countries of Iran and Afghanistan, which report long chains of intermediaries and a heavy reliance on brokers and traders for the collection, sorting and sale of saffron.

A number of studies on saffron distribution channels report that producers achieve only small market shares compared to other players in the distribution chain. Several studies have found that producers receive only a fraction of the purchase price of a product and, in some cases, significantly less than the purchase price, with the majority going to intermediaries as profit, as most producers have no information about product prices at the point of sale. Particular attention was paid to diagnostic tests and country reports, which indicate that saffron markets are generally unorganized and lack adequate regulatory structures. This lack of institutions is reflected in the absence of suitable platforms for public pricing, appropriate structures for enforcing quality grades and standards, and suitable exchange markets for saffron. As a result, price transparency in the agricultural sector is low, as farmers have no knowledge of local, national and export prices or price differences for different qualities of saffron. Since transactions, including all

Table 6.1. World saffron exports and imports (2016–2020, USD thousand) (Karbasi and Zandi Dareh Gharibi, 2022).

Country/Region	Export 2016	Export 2017	Export 2018	Export 2019	Export 2020	Import 2016	Import 2017	Import 2018	Import 2019	Import 2020
World total	403,561	446,327	471,555	213,898	214,164	229,832	261,746	301,023	222,461	218,226
Iran	286,046	325,650	351,127	103,119	90,078	–	–	–	–	–
Spain	65,675	60,114	55,755	47,734	48,464	52,954	54,370	35,783	32,626	29,916
Afghanistan	3,869	9,797	21,235	26,416	45,637	–	–	–	–	–
Greece	3,187	4,185	4,343	3,131	4,212	–	–	–	–	–
France	6,584	7,421	4,732	4,487	3,744	13,323	11,495	9,850	7,699	7,452
Portugal	10,905	7,300	4,355	3,780	3,181	3,929	6,326	3,505	4,902	4,721
Poland	46	966	3,108	2,417	2,477	–	–	–	–	–
Hong Kong, China	5,208	5,487	6,203	5,444	2,406	8,419	20,581	99,683	33,990	35,951
United Arab Emirates (UAE)	2,143	7,839	2,340	2,062	1,838	16,665	23,547	21,505	18,247	4,302
Netherlands	4,311	4,558	3,454	4,145	1,774	–	–	–	–	–
India	1,386	1,547	835	997	1,650	14,912	16,576	18,418	18,354	23,019
USA	–	–	–	–	–	14,944	16,417	16,560	15,761	14,846
China	1,000	500	1,000	500	1,000	3,526	14,516	3,680	2,468	11,153
Germany	2,053	1,975	2,878	1,992	1,359	4,485	4,873	5,391	4,830	3,756
Italy	911	996	1,267	1,029	1,021	17,441	16,769	12,958	11,104	10,553
Sweden	–	–	–	–	–	11,088	9,090	7,022	6,545	7,139
Switzerland	–	–	–	–	–	5,292	4,606	3,786	3,473	3,426

A dash (–) indicates no significant reported trade data for that year in the corresponding category; values are in thousand USD. This table allows a quick comparison of exports vs. imports for major countries and regions, highlighting Iran's dominant share in exports but minimal imports. India, the UAE, Hong Kong and Italy are significant importers despite small production trends in rising Afghan exports and Hong Kong imports.

market functions, are taken care of by brokers and intermediaries, farmers find themselves price-takers with little incentive to capture value addition in processing, traceability and certification. This has important implications for livelihoods in the countryside as well as upgrading value chains. Since producers are not able to capture value, it becomes difficult for them to invest in enhanced post-harvest practices for access to valuable markets in international trade. Numerous value chain projects with interventions for improving market information for farmer groups have been able to increase producer shares (Azimy *et al.*, 2020).

6.4 Strategies to Improve Transparency and Increase Producer Margins

The literature and practice point to a range of different measures that complement each other to improve pricing and producer prices in saffron value chains. For example, producer cooperatives and companies can help to bundle their goods, negotiate better terms and take on some value-adding activities themselves, such as sorting and initial processing. Pilot projects in Afghanistan and Iran have already demonstrated the positive impact of producer organizations. In addition, publicly available wholesale and producer price lists (nationwide) help to eliminate information asymmetries and have therefore been proposed in diagnostic studies for their significant role in improving transparency.

The development of traceable classification systems using objective criteria such as saffron/crocin content, ISO, AOAC International and so on, and a classified quality with a correspondingly higher market price, will encourage producers to invest in appropriate cultivation and processing methods in order to benefit from additional income from better saffron quality. Reliable classification systems will also improve the efficiency of saffron processing for all members of the value chain. Related studies support this argument, claiming that the development of a traceability mechanism that assesses saffron quality and thus influences customer purchasing behaviour has a major positive impact on saffron producers to invest in cultivation methods that meet higher standards. Institutional development strategies in saffron value chains in line with UNIDO recommendations are important to further improve the success of small producers in value chains. These proposals are contained in a report by the World Bank and UNIDO on value chain development in Afghanistan, which highlights the positive impact of value chain development on male and female producers in order to increase their benefits in Afghanistan's agricultural value chains. Pre-harvest financing linked to quality standards and contracts with processors/exporters based on a public list of terms will help producers achieve better market returns while avoiding sales to opportunistic middlemen (World Bank, 2011; UNIDO, 2014).

6.5 Minimizing Raw Material Exports and Promoting Value-added Saffron Processing

One of the most pressing issues in the saffron industry, which directly affects the cost efficiency and economic gains of nations and organizations operating in this sector, concerns the development and technological improvement of saffron cultivation. The use of improved cultivation methods not only maximizes yields but also increases the phytochemical value of saffron. Processing saffron into medicinal concentrates, dietary supplements and high-quality formulations, particularly through nanotechnology, offers enormous economic potential, as producers could then tap into niche markets around the world. In addition, saffron by-products such as petals, stamens and residues from the bulb offer further sources of income. Importantly, a large number of companies benefiting from this technology can generate profits comparable to or even exceeding those from saffron exports by developing innovative saffron-based products.

6.6 International Quality Standards for Saffron

The standardization of saffron has been established in relation to a number of important systems, with ISO 3632 playing a dominant role over all other systems and serving as the international standard for classifying saffron of different qualities into different grades (wholesale, foreign and inferior saffron), whereby the most important bioactive components of saffron, such as crocin, picrocrocins and safranal, are analysed spectrophotometrically. In addition to this standard, the FAO-WHO Codex Alimentarius contains essential food hygiene requirements, such as specific limits for contaminants, pesticides, microbiological counts and heavy metals, with which saffron must comply. The European Union (EU) has some of the strictest regulations in the world regarding pesticide residues, hygiene, labelling and traceability. Additional protection is provided by protected designations of origin such as 'La Mancha' in Spain, 'Kozani' in Greece and 'L'Aquila' in Italy. The main saffron-producing countries, such as Iran, India, Spain, Afghanistan and Morocco, rely on their own standards, including ISIRI, AGMARK, UNE and other national regulations, which are based on ISO standards.

Furthermore, in most cases, saffron has been found to be regularly mixed with safflower, turmeric, marigold, artificial colourings and other fillers. Although modern analyses increasingly rely on HPLC, GC-MS, FTIR and DNA barcoding, relatively new approaches such as NMR spectroscopy and Raman imaging in conjunction with chemometric analyses promise efficient approaches for rapid and reliable verification of saffron quality. At the same time, efforts are being made to develop a digital traceability concept using blockchain, QR code-based authentication, GPS-based mapping of farms and document transparency in order to strengthen customer confidence and effectively combat counterfeiting. Packaging and storage are also important aspects for saffron in order to preserve the product. This includes airtight and UV-protected containers, a humidity-controlled environment and a dark environment for a period of 2–4 years. From this perspective, saffron has a future if the focus is on meeting the growing global demand for common standards in the integration of chemical analysis, sensory analysis and sustainability, the application of AI technology for authentication, further advances in traceability, and a focus on ethical working practices and sustainable farming methods.

In conclusion, saffron cultivation remains an economically viable enterprise, as evidenced by substantial net returns driven by high market prices and by-product sales (Mehmeti *et al.*, 2024). To ensure its sustainable growth, efforts should focus on reducing supply chain costs, enhancing regional employment and economic development, and improving saffron quality, particularly through organic production methods. While traditional agricultural knowledge plays a crucial role in cultivation and processing, it should be preserved, documented and thoughtfully integrated with emerging technologies. Additionally, the socio-cultural significance of saffron as a family-based crop must continue to be valued. Promising new opportunities, such as valorizing saffron waste for cosmetics and fragrances, along with adopting e-commerce and innovative cultivation technologies, can further strengthen the supply chain and improve the livelihoods of saffron farmers (Khajeh-Hosseini and Fallahpour, 2020).

References

Abarghooie, H., Gholavand, A., Mazaheri, D., Noormohammadi, G.H. and Sanaeie, M. (1999) The effect of temperature on the flowering and potential yield of Iranian saffron accessions. *Iranian Journal of Structural Research* 49, 65–69.

Abdullaev, F.I. (2002) Cancer chemopreventive and tumoricidal properties of saffron (*Crocus sativus* L.). *Experimental Biology and Medicine* 227(1), 20–25. DOI: 10.1177/153537020222700104kzz4.

Abe, K. and Saito, H. (2000) Effects of saffron extract and its constituent crocin on learning behaviour and long-term potentiation. *Phytotherapy Research* 14(3), 149–152. DOI: 10.1002/(sici)1099-1573(200005)14:3<149::aid-ptr665>3.0.co;2-5.

Abirafeh, L. (2009) *Gender and International Aid in Afghanistan: A case for Women's Empowerment*. McFarland, Jefferson, NC, pp. 142–145.

Abrishami, M.H. (1997) *Persian Saffron: A Comprehensive Cultural and Agricultural History*. Astan Ghods Razavi Publication, Mashhad.

Acharya, B.S., Ghalehgolabbehbahani, A., Hamido, S., Zinati, G., Bozzolo, A. *et al.* (2025) Saffron (*Crocus sativus* L.): The golden spice — management, challenges, and opportunities for sustainable production in the United States. *Journal of Agriculture and Food Research* 21, 101970. DOI: 10.1016/j.jafr.2025.101970.

Achaya, K.T. (1994) *Indian Food: A Historical Companion*. Oxford University Press, Delhi, pp. 175–178.

Advay, M. (2024) A new crocus species from series *Adamii* from western Iran. *International Rock Gardener* 172, May.

Agha-Hosseini, A.T., Kashani, L., Aleyaseen, A., Ghoreishi, A., Rahmanpour, H.A.L.E.H. *et al.* (2008) Saffron (*Crocus sativus*) for treatment of premenstrual syndrome: A double-blind, randomized and placebo-controlled trial. *BJOG: An International Journal of Obstetrics & Gynaecology* 115(4), 515–519. DOI: 10.1111/j.1471-0528.2007.01652.x.

Ahhund-Zade, I.M. and Muzaferova, R.S. (1975) Radiobiological effect of gamma radiation on *Crocus sativus* L. *Radiobiologiya* 15(2), 319–322.

Ahmad, M. and Sagar, V. (2007) Annual progress report on integrated management of corm/tuber rot of saffron and kala-zeera. In: *Horticulture Technology Mini Mission-1 Research Project No*. Indian Council for Agricultural Research (ICAR), India.

Ahmad, M., Zaffar, G., Mir, S.D., Razvi, S.M., Rather, M.A. *et al.* (2011) Saffron (*Crocus sativus* L.) strategies for enhancing productivity. *Research Journal of Medicinal Plant* 5(6), 630–649. DOI: 10.3923/rjmp.2011.630.649.

Ahooran, M., Hosseini, R. and Zarghami, R. (2009) Plantlet regeneration from saffron protoplast. In: *Sixth Iranian Horticultural Science Congress*, pp. 13–16.

Ahrazem, O., Rubio-Moraga, A., Nebauer, S.G., Molina, R.V. and Gómez-Gómez, L. (2015) Saffron: Its phytochemistry, developmental processes, and biotechnological prospects. *Journal of Agricultural and Food Chemistry* 63(40), 8751–8764. DOI: 10.1021/acs.jafc.5b03194.

Ahrazem, O., Rubio-Moraga, A., Berman, J., Capell, T., Christou, P. *et al.* (2016) The carotenoid cleavage dioxygenase CCD2 catalysing the synthesis of crocetin in spring crocuses and saffron is a plastidial enzyme. *The New Phytologist* 209(2), 650–663. DOI: 10.1111/nph.13609.

Akhondzadeh, S., Fallah-Pour, H., Afkham, K., Jamshidi, A.-H. and Khalighi-Cigaroudi, F. (2004) Comparison of *Crocus sativus* L. and imipramine in the treatment of mild to moderate depression: A pilot double-blind randomized trial [ISRCTN45683816]. *BMC Complementary and Alternative Medicine* 4, 12. DOI: 10.1186/1472-6882-4-12.

Akhondzadeh, S., Sabet, M.S., Harirchian, M.H., Togha, M., Cheraghmakani, H. *et al.* (2010a) Saffron in the treatment of patients with mild to moderate Alzheimer's disease: A 16-week, randomized and placebo-controlled trial. *Journal of Clinical Pharmacy and Therapeutics* 35(5), 581–588. DOI: 10.1111/j.1365-2710.2009.01133.x.

Akhondzadeh, S., Shafiee Sabet, M., Harirchian, M.H., Togha, M., Cheraghmakani, H. *et al.* (2010b) A 22-week, multicenter, randomized, double-blind controlled trial of *Crocus sativus* in the treatment of mild-to-moderate Alzheimer's disease. *Psychopharmacology* 207(4), 637–643. DOI: 10.1007/s00213-009-1706-1.

Alavi-Kia, S.S., Mohammadi, S.A., Aharizad, S. and Moghaddam, M. (2008) Analysis of genetic diversity and phylogenetic relationships in *Crocus genus* of Iran using inter-retrotransposon amplified polymorphism. *Biotechnology & Biotechnological Equipment* 22(3), 795–800. DOI: 10.1080/13102818.2008.10817555.

Alighaleh, P., Pakdel, R., Ghanei Ghooshkhaneh, N., Einafshar, S., Rohani, A. *et al.* (2023) Detection and classification of saffron adulterants by Vis-NIR imaging, chemical analysis, and soft computing. *Foods* 12(11), 2192. DOI: 10.3390/foods12112192.

Alizadah, M.H., Hosainy, S.A.H. and Ghafori, S. (2026) Production and export performance of Afghan saffron over the past decade. *Journal of Natural Science Review* 4(1), 114–126. DOI: 10.62810/jnsr.v4i1.387.

Alonso, G.L., Zalacain, A. and Carmona, M. (2012) Saffron. In: Peter, K.V. (ed.) *Handbook of Herbs and Spices*. Woodhead Publishing, Cambridge, UK, pp. 469–498.

Alsayied, N.F., Fernández, J.A., Schwarzacher, T. and Heslop-Harrison, J.S. (2015) Diversity and relationships of *Crocus sativus* and its relatives analysed by inter-retroelement amplified polymorphism (IRAP). *Annals of Botany* 116(3), 359–368. DOI: 10.1093/aob/mcv103.

Anastasaki, E., Kanakis, C., Pappas, C., Maggi, L., del Campo, C.P. *et al.* (2010) Differentiation of saffron from four countries by mid-infrared spectroscopy and multivariate analysis. *European Food Research and Technology* 230(4), 571–577. DOI: 10.1007/s00217-009-1197-7.

Angeli, L., Muñoz, E.C., Ballabio, D., Morozova, K. and Scampicchio, M. (2026) 1H NMR spectroscopy combined with chemometrics for detection of turmeric adulteration in Italian saffron (*Crocus sativus* L.). *Food Control* 179, 111560. DOI: 10.1016/j.foodcont.2025.111560.

Asgari, M., Yu, Q., Abdi, M., Du, G.L. and Shen, Y.H. (2022) Survey of the history and applications of saffron. *Chinese Medicine and Culture* 5(1), 31–38. DOI: 10.1097/MC9.0000000000000006.

Asgarpanah, M.M. and Kazemivash, R. (2013) Saffron and its derivatives for metabolic syndrome. *Journal of Endocrinological Investigation* 36(10), 857–864.

Ashrafian, S., Zarrineh, M., Jensen, P., Nawrocki, A., Rezadoost, H. *et al.* (2022) Quantitative phosphoproteomics and acetylomics of safranal anticancer effects in triple-negative breast cancer cells. *Journal of Proteome Research* 21(11), 2566–2585. DOI: 10.1021/acs.jproteome.2c00168.

Azimy, M.W., Khan, G.D., Yoshida, Y. and Kawata, K. (2020) Measuring the impacts of saffron production promotion measures on farmers' policy acceptance probability: A randomized conjoint field experiment in Herat Province, Afghanistan. *Sustainability* 12(10), 4026. DOI: 10.3390/su12104026.

Bahrami, B. (2020) The historical dispersal of saffron: How Persian caravans shaped global trade. *Journal of Ethnic Foods* 7(1), 12. DOI: 10.1186/s42779-020-00047-3.

Bathaie, S.Z. and Mousavi, S.Z. (2010) New applications and mechanisms of action of saffron and its important ingredients. *Critical Reviews in Food Science and Nutrition* 50(8), 761–786. DOI: 10.1080/10408390902773003.

Batmanglij, N. (2011) *Food of Life: Ancient Persian and Modern Iranian Cooking and Ceremonies*, 25th anniversary edn. Mage Publishers, Washington, DC, pp. 456–458.

Behzadi, H. and Shahsavand, S. (2009) Historical and cultural roots of saffron in Persia. *Journal of Middle Eastern Studies* 45, 113–129.

Bergomi, A., Comite, V., Santagostini, L., Guglielmi, V. and Fermo, P. (2022) Determination of saffron quality through a multi-analytical approach. *Foods* 11(20), 3227. DOI: 10.3390/foods11203227.

Bhooma, V., Vassou, S.L., Kaliappan, I. and Parani, M. (2023) Identification of adulteration in the market samples of saffron using morphology, HPLC, HPTLC, and DNA barcoding methods. *Genome* 67(2), 43–52. DOI: 10.1139/gen-2022-0059.

Bisti, S., Maccarone, R. and Falsini, B. (2014) Saffron and retina: Neuroprotection and pharmacokinetics. *Visual Neuroscience* 31(4–5), 355–361. DOI: 10.1017/S0952523814000108.

Bononi, M., Milella, P. and Tateo, F. (2015) Gas chromatography of safranal as preferable method for the commercial grading of saffron (*Crocus sativus* L.). *Food Chemistry* 176, 17–21. DOI: 10.1016/j.foodchem.2014.12.047.

Bowen, S. (2015) *Divided Spirits: Tequila, Mezcal, and the Politics of Production*. University of California Press, Berkeley, CA, pp. 33–35.

Broadhead, G.K., Grigg, J.R., McCluskey, P., Hong, T., Schlub, T.E. *et al.* (2012) The neuroprotective effects of crocetin after focal cerebral ischemia in rats. *Journal of Neuroscience Research* 90(7), 1420–1428.

Broadhead, G.K., Grigg, J.R., McCluskey, P., Hong, T., Schlub, T.E. *et al.* (2019) Saffron therapy for the treatment of mild/moderate age-related macular degeneration: A randomised clinical trial. *Graefe's Archive for Clinical and Experimental Ophthalmology* 257(1), 31–40. DOI: 10.1007/s00417-018-4163-x.

Brockington, J. (2019) Symbolism of the saffron robe: Continuity and change in Hindu ascetic dress. In: Joshi, A. and Basu, R. (eds) *Dress and Identity in South Asia*. Bloomsbury Academic, London, pp. 45–62.

Busconi, M., Colli, L., Sánchez, R.A., Santaella, M., De-Los-Mozos Pascual, M. *et al.* (2015) AFLP and MS-AFLP analysis of the variation within saffron crocus (*Crocus sativus* L.) germplasm. *PloS ONE* 10(4), e0123434. DOI: 10.1371/journal.pone.0123434.

Byrd, W.A. (2012) *Economic Development and State-Building in Afghanistan*. USIP Special Report.

Caballero-Ortega, H., Pereda-Miranda, R. and Abdullaev, F.I. (2007) HPLC quantification of major active components from 11 different saffron (*Crocus sativus* L.) sources. *Food Chemistry* 100(3), 1126–1131. DOI: 10.1016/j.foodchem.2005.11.020.

Caiola, M.G. (1999) Reproduction biology of saffron and its allies. In: Negbi, M. (ed.) *Saffron*. CRC Press, London, pp. 27–40. DOI: 10.1201/9780203303665-9.

Caiola, M.G. (2005) Embryo origin and development in *Crocus sativus* L. (Iridaceae). *Plant Biosystems – An International Journal Dealing with All Aspects of Plant Biology* 139(3), 335–343. DOI: 10.1080/11263500500340763.

Caiola, M.G., Somma, D. and Lauretti, P. (2000) Comparative study of pollen and pistil in *Crocus sativus* L. (Iridaceae) and allied species. *Annali Di Botanica* 58.

Cardone, L., Castronuovo, D., Perniola, M., Cicco, N. and Candido, V. (2020) Saffron (*Crocus sativus* L.), the king of spices: An overview. *Scientia Horticulturae* 272, 109560. DOI: 10.1016/j.scienta.2020.109560.

Carmona, M., Zalacain, A., Sánchez, A.M., Novella, J.L. and Alonso, G.L. (2006) Crocetin esters, picrocrocin and its related compounds present in *Crocus sativus* stigmas and *Gardenia jasminoides* fruits. Tentative identification of seven new compounds by LC-ESI-MS. *Journal of Agricultural and Food Chemistry* 54(3), 973–979. DOI: 10.1021/jf052297w.

Caser, M., Victorino, Í.M.M., Demasi, S., Berruti, A., Donno, D. *et al.* (2018) Saffron cultivation in marginal Alpine environments: How AMF inoculation modulates yield and bioactive compounds. *Agronomy* 9(1), 12. DOI: 10.3390/agronomy9010012.

Caser, M., Demasi, S., Victorino, Í.M.M., Donno, D., Faccio, A. *et al.* (2019) Arbuscular mycorrhizal fungi modulate the crop performance and metabolic profile of saffron in soilless cultivation. *Agronomy* 9(5), 232. DOI: 10.3390/agronomy9050232.

Chamkhi, I., Sbabou, L. and Aurag, J. (2023) Improved growth and quality of saffron (*Crocus sativus* L.) in the field conditions through inoculation with selected native plant growth-promoting rhizobacteria (PGPR). *Industrial Crops and Products* 197, 116606. DOI: 10.1016/j.indcrop.2023.116606.

Chen, J. (2000) Occurrence and control of mosaic disease [turnip mosaic virus] in saffron (*Crocus sativus*). *Zhejiang Nongye Kexue* 3, 132–135.

Chen, Y., Zhao, C. and Zhang, H. (2010) Protective effect of crocetin against cerebral ischemia/reperfusion injury in rats. *Chinese Pharmacological Bulletin* 26(11), 1484–1488.

Chib, S., Thangaraj, A., Kaul, S., Dhar, M.K. and Kaul, T. (2020) Development of a system for efficient callus production, somatic embryogenesis and gene editing using CRISPR/Cas9 in saffron (*Crocus sativus* L.). *Plant Methods* 16, 1–10. DOI: 10.1186/s13007-020-00589-2.

Chichiriccò, G. (1999) *Sterility and Perspectives for Genetic Improvement of Crocus sativus.* Harwood Academic Publishers, Amsterdam, The Netherlands, pp. 127–135.

Chichiriccò, G. and Caiola, M.G. (1982) Germination and viability of the pollen of *Crocus sativus* L. *Plant Biosystem* 116(3–4), 167–173. DOI: 10.1080/11263508209428062.

Chourak, Y., Belarbi, E.H., Martínez Rivera, E.Y., da Cunha-Chiamolera, T.P.L., Peña-Fernández, A.A. *et al.* (2021) Fertigation temperature adjustment enhances the yield and quality of saffron grown in a soilless culture system. *HortScience* 56(10), 1191–1194. DOI: 10.21273/HORTSCI16005-21.

Cid-Pérez, T.S., Nevárez-Moorillón, G.V., Ochoa-Velasco, C.E., Navarro-Cruz, A.R., Hernández-Carranza, P. *et al.* (2021) The relation between drying conditions and the development of volatile compounds in saffron (*Crocus sativus*). *Molecules* 26(22), 6954. DOI: 10.3390/molecules26226954.

Cincotta, R.P. (2015) Alternative development in Afghanistan: Saffron as a substitute crop. *Journal of International Development* 27(6), 897–912.

Condurso, C., Cincotta, F., Tripodi, G. and Verzera, A. (2016) Bioactive volatiles in Sicilian (South Italy) saffron: Safranal and its related compounds. *Journal of Essential Oil Research* 29, 221–227.

Consonni, R., Ordoudi, S.A., Cagliani, L.R., Tsiangali, M. and Tsimidou, M.Z. (2016) On the traceability of commercial saffron samples using ^{1}H-NMR and FT-IR metabolomics. *Molecules* 21(3), 286. DOI: 10.3390/molecules21030286.

Dadkhah, M.R., Ehtesham, M. and Fekrat, H. (2003) *Iranian Saffron: An Unknown Jewel.* Shahr Ashoob Publication, Tehran, pp. 1–20.

Dai, R.C., Nabil, W.N.N. and Xu, H.X. (2021) The history of saffron in China: From its origin to applications. *Chinese Medicine and Culture* 4(4), 228–234. DOI: 10.4103/CMAC.CMAC_38_21.

Dalby, A. (2000) *Dangerous Tastes: The Story of Spices.* University of California Press, Berkeley, CA.

Dar, M.H., Groach, R., Razvi, S.M. and Singh, N. (2017) Saffron crop (golden crop) in modern sustainable agricultural systems. *International Journal for Research in Applied Science and Engineering Technology* V(XI), 247–259. DOI: 10.22214/ijraset.2017.11037.

Daruwalla, A. and Kiser, P.D. (2020) Structural and mechanistic aspects of carotenoid cleavage dioxygenases (CCDs). *Biochimica et Biophysica Acta (BBA) - Molecular and Cell Biology of Lipids* 1865(11), 158590. DOI: 10.1016/j.bbalip.2019.158590.

Darvishi, E., Zarghami, R., Mishani, C.A., Omidi, M. and Sarkhosh, A. (2006) *In vitro* production of pathogen-free plantlets via meristem culture in saffron (*Crocus sativus* L.). *Biotechnology* 5(3), 292–295. DOI: 10.3923/biotech.2006.292.295.

Dastranj, M., Sepaskhah, A.R. and Kamgar-Haghighi, A.A. (2019) Rainfall and its distribution influences on rain-fed saffron yield and economic analysis. *Theoretical and Applied Climatology* 137(3–4), 3139–3147. DOI: 10.1007/s00704-019-02804-0.

Davis, J.M. (2011) Saffron (*Crocus sativus* L.): An ethnobotanical overview. *HerbalGram* 92, 1–12.

Day, J. (2011a) Counting threads: Saffron in Aegean Bronze Age writing and society. *Oxford Journal of Archaeology* 30(4), 369–391. DOI: 10.1111/j.1468-0092.2011.00372.x.

Day, J. (2011b) Crocuses in context: A diachronic survey of the crocus motif in the Aegean Bronze Age. *Hesperia* 80(3), 337. DOI: 10.2972/hesperia.80.3.0337.

del Campo, C.P., Carmona, M., Maggi, L., Kanakis, C.D., Anastasaki, E.G. *et al.* (2010) Picrocrocin content and quality categories in different (345) worldwide samples of saffron (*Crocus sativus* L.). *Journal of Agricultural and Food Chemistry* 58(2), 1305–1312. DOI: 10.1021/jf903336t.

Delgado, M.C., Aramburu, A.Z. and Díaz-Marta, G.L.A. (2006) *The Chemical Composition of Saffron: Color, Taste and Aroma*. Bomarzo, Albacete, Spain.

Demurtas, O.C., Diretto, G., Sulli, M., Aprea, G., Giuliano, G. *et al.* (2016) Functional characterization of a carotenoid cleavage dioxygenase from saffron (*Crocus sativus* L.) involved in the biosynthesis of crocetin. *Acta Horticulturae* 1184, 251–258. DOI: 10.17660/ActaHortic.2017.1184.35.

Demurtas, O.C., Frusciante, S., Ferrante, P., Diretto, G., Azad, N.H. *et al.* (2018) Candidate enzymes for saffron crocin biosynthesis are localized in multiple cellular compartments. *Plant Physiology* 177(3), 990–1006. DOI: 10.1104/pp.17.01815.

Demurtas, Olivia Costantina, de Brito Francisco, R., Diretto, G., Ferrante, P., Frusciante, S. *et al.* (2019) ABCC transporters mediate the vacuolar accumulation of crocins in saffron stigmas. *The Plant Cell* 31(11), 2789–2804. DOI: 10.1105/tpc.19.00193.

Deo, B. (2003) Growing saffron – The world's most expensive spice. *Crop Food Research* 20(1), 1–4.

Dewan, R. (2015) Bronze Age flower power: The Minoan use and social significance of saffron and crocus flowers. *Chronica* 5, 42–55.

Dhar, M.K., Sharma, M., Bhat, A., Chrungoo, N.K. and Kaul, S. (2017) Functional genomics of apocarotenoids in saffron: Insights from chemistry, molecular biology and therapeutic applications. *Briefings in Functional Genomics* 16(6), 336–347. DOI: 10.1093/bfgp/elx003.

Diretto, G., Ahrazem, O., Rubio-Moraga, Á., Fiore, A., Sevi, F. *et al.* (2019) UGT709G1: A novel uridine diphosphate glycosyltransferase involved in the biosynthesis of picrocrocin, the precursor of safranal in saffron (*Crocus sativus*). *The New Phytologist* 224(2), 725–740. DOI: 10.1111/nph.16079.

Dolatyari, A., Tolyat Abolhasani, M., Ardalani, F. and Rukšāns, J. (2024) A taxonomic revision of the genus *Crocus* (Iridaceae) in Iran. *Nordic Journal of Botany* 2024(6), e04270. DOI: 10.1111/njb.04270.

Dole, J.M. (2003) Research approaches for determining cold requirements for forcing and flowering of geophytes. *HortScience* 38(3), 341–346. DOI: 10.21273/HORTSCI.38.3.341.

D'Onofrio, G., Nabavi, S.M., Sancarlo, D., Greco, A. and Pieretti, S. (2021) *Crocus sativus* L. (saffron) in Alzheimer's disease treatment: Bioactive effects on cognitive impairment. *Current Neuropharmacology* 19(9), 1606–1616. DOI: 10.2174/1570159X19666210113144703.

Douglas, M.H., Smallfield, B.M., Wallace, A.R. and McGimpsey, J.A. (2014) Saffron (*Crocus sativus* L.): The effect of mother corm size on progeny multiplication, flower and stigma production. *Scientia Horticulturae* 166, 50–58. DOI: 10.1016/j.scienta.2013.12.007.

Dowlatabadi, R., Farshidfar, F., Zare, Z., Pirali, M., Rabiei, M. *et al.* (2017) Detection of adulteration in Iranian saffron samples by ^{1}H NMR spectroscopy and multivariate data analysis techniques. *Metabolomics* 13(2), 19. DOI: 10.1007/s11306-016-1155-x.

El-Beshbishy, H.A., Hassan, M.H., Aly, H.A.A., Doghish, A.S. and Alghaithy, A.A.A. (2012) Crocin protects against cognitive deficits and oxidative stress induced by beryllium chloride in rats. *European Review for Medical and Pharmacological Sciences* 16(7), 960–970.

Elliott, S.J. (2017) Climate suitability for saffron cultivation in Australia. *Agricultural Systems* 152, 44–52.

El-Maraghy, S.A., Rizk, S.M. and Shahin, N.N. (2009) Protective effect of crocin against gamma radiation-induced oxidative stress and histopathological changes in the brain of rats. *International Journal of Radiation Biology* 85(11), 1079–1088.

Erol, O., Kaya, H.B., Şik, L., Tuna, M., Can, L. *et al.* (2014) The genus *Crocus*, series *Crocus* (Iridaceae) in Turkey and 2 East Aegean islands: A genetic approach. *Turkish Journal of Biology* 38(1), 48–62. DOI: 10.3906/biy-1305-14.

European Commission (2020) Protected Designation of Origin (PDO): Azafrán de La Mancha. European Commission, Agriculture and Rural Development, Brussels. Available at: https://ec.europa.eu/info/food-farming-fisheries/food-safety-and-quality/certification/quality-labels/geographical-indications-register/ (accessed 21 January 2026).

Fallahi, H.R., Zamani, G., Aghhavani-Shajari, M. and Samadzadeh, A. (2017) Comparison of flowering and growth of saffron in natural and controlled culture systems. In: *Proceedings of the 6th National Congress of Medicinal Plants*, Tehran (Iran), May, pp. 9–10.

Fallahi, H.R., Abbasi Aval Bohlooli, S., Noferesti, E., Hoseini, S.M., Seddigh Makoo, S. *et al.* (2020) Evaluation of the possibility of saffron transplanting and corm production in soil-less planting system. *Journal of Saffron Research* 8(2), 269–284.

FAO (Food and Agriculture Association of the United Nations) (2022) FAOSTAT: Saffron production statistics. Available at: https://www.fao.org/faostat/ (accessed 10 September 2025)

Farag, M.A., Hegazi, N., Dokhalahy, E. and Khattab, A.R. (2020) Chemometrics based GC-MS aroma profiling for revealing freshness, origin and roasting indices in saffron spice and its adulteration. *Food Chemistry* 331, 127358. DOI: 10.1016/j.foodchem.2020.127358.

Farooqi, A.A. and Sreeramu, B.S. (2004) *Cultivation of Medicinal and Aromatic Crops*. Universities Press, Hyderabad, India.

Fernández, J.A. (2004) Biology, biotechnology and biomedicine of saffron. In: Negbi, M. (ed.) *Saffron: Crocus Sativus* L. Harwood Academic Publishers, Amsterdam, The Netherlands, pp. 127–159.

Fernández, J.-A., Santana, O., Guardiola, J.-L., Molina, R.-V., Heslop-Harrison, P. *et al.* (2011) The world saffron and *Crocus* collection: Strategies for establishment, management, characterisation and utilisation. *Genetic Resources and Crop Evolution* 58(1), 125–137. DOI: 10.1007/s10722-010-9601-5.

Filatova, M., Hajslová, J. and Stupak, M. (2024) Detection of saffron adulteration by other plant species using SPME-GC-HRMS. *European Food Research and Technology* 250(3), 911–922. DOI: 10.1007/s00217-023-04443-2.

Fiori, M., Ligios, V. and Schiaffino, A. (2011) Identification and characterization of Burkholderia isolates obtained from bacterial rot of saffron (*Crocus sativus* L.) grown in Italy. *Phytopathologia Mediterranea* 50(3), 450–461.

Foschi, M., Tozzi, L., Di Donato, F., Biancolillo, A. and D'Archivio, A.A. (2023) A novel FTIR-based chemometric solution for the assessment of saffron adulteration with non-fresh stigmas. *Molecules* 28(1), 33. DOI: 10.3390/molecules28010033.

Francis, S. and Ramandi, M.T. (2020) *Crocologia – A Detailed Study of Saffron, the King of Plants*. Brill, Leiden, The Netherlands. DOI: 10.1163/9789004435292.

Freedman, P. (2007) *Food: The History of Taste*. University of California Press, Berkeley, CA, pp. 210–212.

Freedman, P. (2008) *Out of the East: Spices and the Medieval Imagination*. Yale University Press, New Haven, CT, pp. 187–190.

Frello, S., Ørgaard, M., Jacobsen, N. and Heslop-Harrison, J.S. (2004) The genomic organization and evolutionary distribution of a tandemly repeated DNA sequence family in the genus *Crocus* (Iridaceae). *Hereditas* 141(1), 81–88. DOI: 10.1111/j.1601-5223.2004.01839.x.

Fujii, S., Morita, Y., Ohta, T., Uto, T. and Shoyama, Y. (2022) Saffron (*Crocus sativus* L.) as a valuable spice and food product: A narrative review. *Longhua Chinese Medicine* 5, 18. DOI: 10.21037/lcm-22-1.

Galigani, P.F. and Pegna, F.G. (1999) Mechanized saffron cultivation, including harvesting. In: Negbi, M. (ed.) *Saffron*. CRC Press, London, pp. 115–126.

Ghadrdoost, B., Vafaei, A.A., Rashidy-Pour, A., Hajisoltani, R., Bandegi, A.R. *et al.* (2011) Protective effects of saffron extract and its active constituent crocin against oxidative stress and spatial learning and memory deficits induced by chronic stress in rats. *European Journal of Pharmacology* 667(1–3), 222–229. DOI: 10.1016/j.ejphar.2011.05.012.

Ghani, M.Y. (2002) Corm rot disease of saffron and its management. In: *Proceedings of Seminar-cum-Workshop on Saffron (Crocus sativus)*, pp. 107–112.

Gheshm, R. and Brown, R.N. (2021) Growing saffron crocus in the northeastern United States: Effects of winter protection and planting density. *HortTechnology* 31(4), 524–531. DOI: 10.21273/HORTTECH04836-21.

Ghorbani, M. (2008) The efficiency of saffron's marketing channel in Iran. *World Applied Sciences Journal* 4(4), 523–527.

Ghorbani, R. and Koocheki, A. (2017) Sustainable cultivation of saffron in Iran. *Sustainable Agriculture Reviews* 169–203. DOI: 10.1007/978-3-319-58679-3_6.

Giorgi, A., Pentimalli, D., Giupponi, L. and Panseri, S. (2017) Quality traits of saffron (*Crocus sativus* L.) produced in the Italian Alps. *Open Agriculture* 2(1), 52–57. DOI: 10.1515/opag-2017-0005.

Gismondi, A., Fanali, F., Labarga, J.M.M., Caiola, M.G. and Canini, A. (2013) *Crocus sativus* L. genomics and different DNA barcode applications. *Plant Systematics and Evolution* 299(10), 1859–1863. DOI: 10.1007/s00606-013-0841-7.

Goliaris, A.H. (1999) Saffron cultivation in Greece. In: Negbi, M. (ed.) *Saffron*. CRC Press, London, pp. 65–76.

Gore, M., Bradbury, P., Hogers, R., Kirst, M., Verstege, E. *et al.* (2007) Evaluation of target preparation methods for single-feature polymorphism detection in large complex plant genomes. *Crop Science* 47(S2), S-135. DOI: 10.2135/cropsci2007.02.0085tpg.

Gregory, M.J., Menary, R.C., and Rural Industries Research and Development Corporation (2007) *Development of an Internationally Competitive Australian Saffron Industry*. Publication No. 07/141. Rural Industries Research and Development Corporation.

Gresta, F., Lombardo, G.M., Siracusa, L. and Ruberto, G. (2008a) Effect of mother corm dimension and sowing time on stigma yield, daughter corms and qualitative aspects of saffron (*Crocus sativus* L.) in a Mediterranean environment. *Journal of the Science of Food and Agriculture* 88(7), 1144–1150. DOI: 10.1002/jsfa.3177.

Gresta, F., Lombardo, G.M., Siracusa, L. and Ruberto, G. (2008b) Saffron, an alternative crop for sustainable agricultural systems. A review. *Agronomy for Sustainable Development* 28(1), 95–112. DOI: 10.1051/agro:2007030.

Gresta, F., Santonoceto, C. and Avola, G. (2016) Crop rotation as an effective strategy for saffron (*Crocus sativus* L.) cultivation. *Scientia Horticulturae* 211, 34–39. DOI: 10.1016/j.scienta.2016.08.007.

Grilli Caiola, M. (2004) Saffron reproductive biology. *Acta Horticulturae 650: I International Symposium on Saffron Biology and Biotechnology* 25–37. DOI: 10.17660/ActaHortic.2004.650.1.

Grilli Caiola, M., Caputo, P. and Zanier, R. (2004) RAPD analysis in *Crocus sativus* L. accessions and related *Crocus* species. *Biologia Plantarum* 48(3), 375–380. DOI: 10.1023/B:BIOP.0000041089.92559.84.

Gupta, V., Sharma, A., Rai, P.K., Gupta, S.K., Singh, B. *et al.* (2021) Corm rot of saffron: Epidemiology and Management. *Agronomy* 11(2), 339. DOI: 10.3390/agronomy11020339.

Hariri, A.T., Moallem, S.A., Mahmoudi, M. and Hosseinzadeh, H. (2010) The effect of crocin and safranal, constituents of saffron, against subacute effect of diazepam on learning and memory in rats. *Phytotherapy Research* 24(5), 726–730.

Harpke, D., Peruzzi, L., Kerndorff, H., Karamplianis, T., Constantinidis, T. *et al.* (2014) Phylogeny, geographic distribution, and new taxonomic circumscription of the *Crocus reticulatus* species group (Iridaceae). *Turkish Journal of Botany* 38, 1182–1198. DOI: 10.3906/bot-1405-60.

Harpke, D., Carta, A., Tomović, G., Ranđelović, V., Ranđelović, N. *et al.* (2015) Phylogeny, karyotype evolution and taxonomy of *Crocus* series Verni (Iridaceae). *Plant Systematics and Evolution* 301(1), 309–325. DOI: 10.1007/s00606-014-1074-0.

Hashemi-Nasab, F.S. and Parastar, H. (2022) Vis-NIR hyperspectral imaging coupled with independent component analysis for saffron authentication. *Food Chemistry* 393, 133450. DOI: 10.1016/j.foodchem.2022.133450.

He, S.Y., Qian, Z.Y., Wen, N., Tang, F.T., Xu, G.L. *et al.* (2005) Protective effects of crocin on acute cerebral ischemia and reperfusion injury in rats. *Chinese Pharmacological Bulletin* 21(6), 730–733.

Heenatigala, Y. (2025) *Quality Assessment of Saffron by UV-Vis Spectroscopy in Accordance with ISO 3632 (Application Note)*. Agilent Technologies, Inc.

Hill, T. (2004) *The Contemporary Encyclopedia of Herbs and Spices: Seasonings for the Global Kitchen*. Wiley, Hoboken, NJ.

Hodgdon, E.S. (2020) Economic potential of saffron production in the United States. *Journal of Agriculture, Food Systems, and Community Development* 10(2), 35–52.

Hosseini, M., Mollafilabi, A. and Nassiri, M.M. (2008) Spatial and temporal patterns in saffron (*Crocus sativus* L.) yield of Khorasan province and their relationship with long term weather variation. *Iranian Journal of Field Crops Research* 6(1), 79–88.

Hosseinzadeh, H. and Ziaee, T. (2006) Effect of *Crocus sativus* L. (saffron) stigma and its constituents, crocin and safranal, on memory and learning in mice. *Journal of Medicinal Plants* 5(18), 40–49.

Hosseinzadeh, H. and Sadeghnia, H.R. (2007) Protective effect of safranal on pentylenetetrazol-induced seizures in the rat: Involvement of GABAergic and opioids systems. *Phytomedicine* 14(4), 256–262. DOI: 10.1016/j.phymed.2006.03.007.

Hosseinzadeh, H. and Ziaee, T. (2008) Effect of *Crocus sativus* L. (saffron) stigma and its constituents, crocin and safranal, on memory and learning in rats performing the Morris water maze task. *Journal of Medicinal Plants* 7(26), 40–48.

Hosseinzadeh, H. and Noraei, N.B. (2009) Anxiolytic and hypnotic effect of *Crocus sativus* aqueous extract and its constituents, crocin and safranal, in mice. *Phytotherapy Research* 23(6), 768–774. DOI: 10.1002/ptr.2597.

Hosseinzadeh, H. and Nassiri-Asl, M. (2013a) Avicenna's (Ibn Sina) the Canon of Medicine and saffron (*Crocus sativus*): A review. *Phytotherapy Research* 27(4), 475–483. DOI: 10.1002/ptr.4784.

Hosseinzadeh, H. and Nassiri-Asl, M. (2013b) Pharmacological effects of *Crocus sativus* and its constituents: A review. *Iranian Journal of Basic Medical Sciences* 16(8), 641–653.

Hosseinzadeh, H., Karimi, G. and Niapoor, M. (2004) Antidepressant effects of *Crocus sativus* stigma extracts and its constituents, crocin and safranal, in mice. *Journal of Medicinal Plants* 3(11), 48–58.

Hosseinzadeh, H., Sadeghnia, H.R., Ziaee, T. and Danaee, A. (2005) Protective effect of aqueous saffron extract (*Crocus sativus* L.) and crocin, its active constituent, on renal ischemia-reperfusion-induced oxidative damage in rats. *Journal of Pharmacy & Pharmaceutical Sciences* 8(3), 387–393.

Hosseinzadeh, H., Ziaee, T. and Sadeghi, A. (2008) The effect of saffron, *Crocus sativus* stigma, extract and its constituents, safranal and crocin on sexual behaviors in normal male rats. *Phytomedicine* 15(6–7), 491–495. DOI: 10.1016/j.phymed.2007.09.020.

Hosseinzadeh, H., Sadeghnia, H.R. and Ziaee, T. (2010) Effect of safranal on extracellular hippocampal levels of glutamate and aspartate during kainic acid treatment in anesthetized rats. *Journal of Ayurveda and Integrative Medicine* 1(4), 223–228.

Hosseinzadeh, H., Sadeghnia, H.R. and Rahimi, A. (2012) Effect of safranal on extracellular hippocampal levels of glutamate and aspartate during kainic acid treatment in anesthetized rats. *Journal of Ayurveda and Integrative Medicine* 3(4), 181–186.

Huang, W., Li, F., Liu, Y. and Long, C. (2014) Identification of *Crocus sativus* (Iridaceae) and its adulterants by using DNA barcoding technique. *Planta Medica* 80(10), PR3. DOI: 10.1055/s-0034-1382739.

Husaini, A.M., Hassan, B., Ghani, M.Y., Silva, J.A. and Kirmani, N.A. (2010) Saffron (*Crocus sativus* Kashmirianus) cultivation in Kashmir: Practices and problems. *Functional Plant Science and Biotechnology* 4(2), 108–115.

Imenshahidi, M., Hosseinzadeh, H. and Javadpour, Y. (2010) The effect of aqueous saffron extract and its constituent, safranal, on morphine withdrawal syndrome in mice. *Phytotherapy Research* 24(5), 726–730. DOI: 10.1002/ptr.3011.

Imenshahidi, M., Ramezani, M., Mousavi, Z. and Hosseinzadeh, H. (2011) Study of the protective effect of saffron extract and crocin on cognitive functions in rats with cerebral ischemia. *Journal of Medicinal Plants* 10(38), 86–97.

International Organization for Standardization (ISO) (1993) *ISO 3632-1:1993: Spices – Saffron (Crocus Sativus* L.) *– Part 1: Specification*. ISO, Geneva, Switzerland.

International Organization for Standardization (ISO) (2010) *ISO 3632-2:2010: Spices – Saffron (Crocus sativus L.) – Part 2: Test Methods*. ISO, Geneva, Switzerland.

International Taste and Quality Institute (iTQi) (2025) Superior taste award for Afghan saffron. Available at: www.itiqi.com (accessed 10 September 2025).

Iqbal, A.M., Lone, A.A., Khan, M.H., uddin Sofi, M., Alie, B.A. *et al.* (2023) Genetic variability assessment of indigenous and exotic saffron germplasm through morpho-agronomic characterization at Jammu and Kashmir, India. *Plant Genetic Resources* 21(5), 389–398. DOI: 10.1017/S1479262123000886.

Izadpanah, F., Kalantari, S., Hassani, M.E., Naghavi, M.R. and Shokrpour, M. (2014) Variation in saffron (*Crocus sativus* L.) accessions and *Crocus* wild species by RAPD analysis. *Plant Systematics and Evolution* 300(8), 1941–1944. DOI: 10.1007/s00606-014-1018-8.

Jadhavar Business Intelligence (2024) *Saffron Market Size, Share, Competitive Landscape and Trend Analysis Report by Product Type, Form, Application and Distribution Channel: Global Opportunity and Industry Forecast (2025–2030)*. Available at: https://www.jadhavarbusinessintelligence.com/market-research-report/saffron-market/1083 (accessed 30 April 2026).

Jafarzadeh, A., Khashei-Siuki, A. and Shahidi, A. (2015) Modeling of climate change effects on saffron water requirement in South Khorasan province by GIS. *Journal of Saffron Research* 3(2), 163–174.

Jarukas, L., Vitkevicius, K., Mykhailenko, O., Bezruk, I., Georgiyants, V. *et al.* (2022) Effective isolation of picrocrocin and crocins from saffron: From HPTLC to working standard obtaining. *Molecules* 27(13), 4286. DOI: 10.3390/molecules27134286.

Kabiri, M., Rezadoost, H. and Ghassempour, A. (2017) A comparative quality study of saffron constituents through HPLC and HPTLC methods followed by isolation of crocins and picrocrocin. *LWT* 84, 1–9. DOI: 10.1016/j.lwt.2017.05.033.

Kafi, M. (ed.) (2006) *Saffron (Crocus sativus): Production and Processing*. CRC Press, Boca Raton, FL. DOI: 10.1201/9781482280463.

Kalha, C.S., Gupta, V., Gupta, D. and Priya, S. (2007) First report of sclerotial rot of saffron caused by *Sclerotium rolfsii* in India. *Plant Disease* 91(9), 1203. DOI: 10.1094/PDIS-91-9-1203B.

Kamalipour, M. and Akhondzadeh, S. (2011) Cardiovascular effects of saffron: An evidence-based review. *The Journal of Tehran Heart Center* 6(2), 59–61.

Karbasi, A. and Zandi Dareh Gharibi, B. (2022) Economic aspects of saffron in the world. In: Vakhlu, J., Ambardar, S., Alireza Salami, S. and Kole, C. (eds) *The Saffron Genome*. Springer International Publishing, Cham, Switzerland, pp. 275–287.

Karimi, H. and Mohammadi, M. (2018) Water scarcity and saffron farming in Iran. *Journal of Arid Environments* 153, 1–9.

Karimi, G., Hosseinzadeh, H. and Khaleghpanah, P. (2004) Study of antidepressant effect of aqueous and ethanolic extract of *Crocus sativus* in mice. *Iranian Journal of Basic Medical Sciences* 7(1), 11–16.

Karp, A., Edwards, K.J., Bruford, M., Funk, S., Vosman, B. *et al.* (1997) Molecular technologies for biodiversity evaluation: Opportunities and challenges. *Nature Biotechnology* 15(7), 625–628. DOI: 10.1038/nbt0797-625.

Kashani, L., Eslatmanesh, S., Saedi, N., Niroomand, N., Ebrahimi, M. *et al.* (2017) Comparison of saffron versus fluoxetine in treatment of mild to moderate postpartum depression: A double-blind, randomized clinical trial. *Pharmacopsychiatry* 50(2), 64–68. DOI: 10.1055/s-0042-115306.

Katouzian, H. (2009) *The Persians: Ancient, Mediaeval and Modern Iran*. Yale University Press, New Haven, CT, pp. 215–217.

Kazemi, F., Vosough, I., Sepahi, S. and Mohajeri, S.A. (2021) Effect of crocin versus fluoxetine in treatment of mild to moderate obsessive-compulsive disorder: A double blind randomized clinical trial. *Human Psychopharmacology* 36(4), e2780. DOI: 10.1002/hup.2780.

Kazemi-Shahandashti, S.-S., Mann, L., El-Nagish, A., Harpke, D., Nemati, Z. *et al.* (2022) Ancient artworks and crocus genetics both support saffron's origin in early Greece. *Frontiers in Plant Science* 13, 834416. DOI: 10.3389/fpls.2022.834416.

Keify, F. and Beiki, A.H. (2012) Exploitation of random amplified polymorphic DNA (RAPD) and sequence-related amplified polymorphism (SRAP) markers for genetic diversity of saffron collection. *Journal of Medicinal Plants Research* 6(14), 2761–2768. DOI: 10.5897/JMPR11.834.

Khajeh-Hosseini, M. and Fallahpour, F. (2020) Emerging innovation in saffron production. In: Koocheki, A. and Khajeh-Hosseini, M. (eds) *Saffron: Science, Technology and Health*. Woodhead Publishing, Cambridge, UK, pp. 205–216. DOI: 10.1016/B978-0-12-818638-1.00012-5.

Khan, I.A. (2004) Induced mutagenic variability in saffron (*Crocus sativus* L.). *Acta Horticulturae 650: I International Symposium on Saffron Biology and Biotechnology* 281–283. DOI: 10.17660/ActaHortic.2004.650.32.

Khan, M.A., Nagoo, S., Naseer, S., Nehvi, F.A. and Zargar, S.M. (2011) Induced mutation as a tool for improving corm multiplication in saffron (*Crocus sativus* L.). *Journal of Phytology* 3(7).

Khilare, V., Tiknaik, A., Prakash, B., Ughade, B., Korhale, G. *et al.* (2019) Multiple tests on saffron find new adulterant materials and reveal that 1st grade saffron is rare in the market. *Food Chemistry* 272, 635–642. DOI: 10.1016/j.foodchem.2018.08.089.

Khorramdel, S., Eskandari Nasrabadi, S. and Mahmoodi, G. (2015) Evaluation of mother corm weights and foliar fertilizer levels on saffron (*Crocus sativus* L.) growth and yield components. *Journal of Applied Research on Medicinal and Aromatic Plants* 2(1), 9–14. DOI: 10.1016/j.jarmap.2015.01.002.

Kiani, S., Yazdanpanah, H. and Feizy, J. (2023) Geographical origin differentiation and quality determination of saffron using a portable hyperspectral imaging system. *Infrared Physics & Technology* 131, 104634. DOI: 10.1016/j.infrared.2023.104634.

Konoshima, T., Takasaki, M., Tokuda, H., Morimoto, S., Tanaka, H. *et al.* (1998) Crocin and crocetin derivatives inhibit skin tumour promotion in mice. *Phytotherapy Research* 12(6), 400–404. DOI: 10.1002/(SICI)1099-1573(199809)12:6<400::AID-PTR321>3.0.CO;2-4.

Koocheki, A.A. (2004) Indigenous knowledge in agriculture with particular reference to saffron production in Iran. *Acta Horticulturae* 650, 175–182. DOI: 10.17660/ActaHortic.2004.650.17.

Koocheki, A. and Seyyedi, S.M. (2016) Effects of corm size, organic fertilizers, Fe-EDTA and Zn-EDTA foliar application on nitrogen and phosphorus uptake of saffron (*Crocus sativus* L.) in a calcareous soil under greenhouse conditions. *Notulae Scientia Biologicae* 8(4), 461–467. DOI: 10.15835/nsb849891.

Koocheki, A. and Khajeh-Hosseini, M. (eds) (2020) *Saffron: Science, Technology and Health*. Woodhead Publishing, Cambridge, UK.

Koocheki, A. and Seyyedi, S.M. (2020) Saffron 'seed', the corm. In: Koocheki, A. and Khajeh-Hosseini, M. (eds) *Saffron: Science, Technology and Health*. Woodhead Publishing, Cambridge, UK, pp. 93–118. DOI: 10.1016/B978-0-12-818638-1.00007-1.

Koocheki, A., Moghaddam, P., Fallahi, H. and Aghhavani-Shajari, M. (2016) The study of saffron (*Crocus sativus* L.): Replacement corms growth in response to planting date, irrigation management and companion crops. *Saffron Agronomy and Technology* 4(1), 3–18. DOI: 10.22048/jsat.2016.11895.

Koocheki, A., Karbasi, A. and Seyyedi, S.M. (2017) Some reasons for saffron yield loss over the last 30 years period. *Saffron Agronomy and Technology* 5(2), 107–122. DOI: 10.22048/jsat.2016.38669.

Kothari, D., Thakur, R. and Kumar, R. (2021) Saffron (*Crocus sativus* L.): Gold of the spices—A comprehensive review. *Horticulture, Environment, and Biotechnology* 62(5), 661–677. DOI: 10.1007/s13580-021-00349-8.

Koulakiotis, N.S., Purhonen, P., Gikas, E., Hebert, H. and Tsarbopoulos, A. (2020) Crocus-derived compounds alter the aggregation pathway of Alzheimer's disease- associated beta amyloid protein. *Scientific Reports* 10(1), 18150. DOI: 10.1038/s41598-020-74770-x.

Kozani Saffron Producers' Cooperative (2026) The cooperative. Kozani Saffron Producers' Cooperative. Available at: https://safran.gr/the-cooperative/ (accessed 29 April 2026).

Krokos Kozanis Producers' Cooperative (2021) Annual Cooperative Report 2021. Krokos Kozanis Producers Cooperative, Kozani, Greece. Available at: https://www.krokoskozanis.gr/en/annual-report-2021 (accessed 21 January 2026).

Kuster, K. (2003) Safran (*Crocus sativus* L.) in der Phytotherapie: Pharmakologie und klinische Anwendung. *Zeitschrift für Phytotherapie* 24(3), 113–122. DOI: 10.1055/s-2003-40474.

Kyriakoudi, A., Ordoudi, S.A., Roldán-Medina, M. and Tsimidou, M.Z. (2015a) Saffron, a functional spice. *Austin Journal of Nutrition and Food Sciences* 3(1), 1059.

Kyriakoudi, A., Tsimidou, M.Z., O'Callaghan, Y.C., Galvin, K. and O'Brien, N.M. (2015b) Changes in total and individual crocetin esters upon *in vitro* gastrointestinal digestion of saffron aqueous extracts. *Journal of Agricultural and Food Chemistry* 61(22), 5318–5327. DOI: 10.1021/jf400540y.

Lachguer, K., Boudadi, I., Lachheb, M., Beraouz, I., El Merzougui, S. *et al.* (2025a) Saffron (*Crocus sativus* L.) cultivation and properties: A review. *International Journal of Horticultural Science and Technology* 12(2), 627–646. DOI: 10.22059/ijhst.2024.377028.855.

Lachguer, K., El Merzougui, S., Boudadi, I., Boukroute, A., Lagram, K. *et al.* (2025b) Innovative green extraction techniques for bioactive compounds from saffron (*Crocus sativus* L.): A focus on crocin and its neuroprotective potential. *Journal of Functional Foods* 124, 105–115.

Lage, M. and Cantrell, C.L. (2009) Quantification of saffron (*Crocus sativus* L.) metabolites crocins, picrocrocin and safranal for quality determination of the spice grown under different environmental Moroccan conditions. *Scientia Horticulturae* 121(3), 366–373. DOI: 10.1016/j.scienta.2009.02.017.

Lage, M., Gaboun, F., Bakhy, K., Dakak, H., Zouahri, A. *et al.* (2010) Sustainable production of high quality saffron (*Crocus sativus* L.) in some Moroccan areas. *Acta Horticulturae 850: III International Symposium on Saffron: Forthcoming Challenges in Cultivation, Research and Economics* 235–238. DOI: 10.17660/ActaHortic.2010.850.39.

Lee, I.A., Lee, J.H., Baek, N.I. and Kim, D.H. (2005) Neuroprotective effects of crocetin against cerebral ischemia/reperfusion injury in rats. *Korean Journal of Pharmacognosy* 36(4), 292–298.

Leone, S., Recinella, A., Chiavaroli, L., Orlando, G., Ferrante, C. *et al.* (2018) Phytotherapic use of the *Crocus sativus* L. (Saffron) and its potential applications: A brief overview. *Phytotherapy Research* 1–12. DOI: 10.1002/ptr.6181.

Lim, T.K. (2012) *Edible Medicinal and Non-Medicinal Plants*, Vol. 1. Springer, Dordrecht, The Netherlands, pp. 285–292. DOI: 10.1007/978-90-481-8661-7_43.

Lin, J.K. and Wang, C.J. (1986) Protection of crocin dyes on the acute hepatic damage induced by aflatoxin B1 and dimethylnitrosamine in rats. *Carcinogenesis* 7(4), 595–599. DOI: 10.1093/carcin/7.4.595.

Liu, J.X. and Qian, Z.Y. (2005) Protective effects of crocetin on acute cerebral ischemia in rats. *Chinese Pharmacological Bulletin* 21(5), 578–581.

López-Jimenez, A.J., Frusciante, S., Niza, E., Ahrazem, O., Rubio-Moraga, Á. *et al.* (2021) A new glycosyltransferase enzyme from family 91, UGT91P3, is responsible for the final glucosylation step of crocins in saffron (*Crocus sativus* L.). *International Journal of Molecular Sciences* 22(16), 8815. DOI: 10.3390/ijms22168815.

López-Sampson, A. and Page, T. (2018) History of use and trade of agarwood. *Economic Botany* 72(1), 107–129. DOI: 10.1007/s12231-018-9408-4.

Lu, Y., Zhang, C., Feng, K., Luan, J., Cao, Y. *et al.* (2024) Characterization of saffron from different origins by HS-GC-IMS and authenticity identification combined with deep learning. *Food Chemistry* 24, 101981. DOI: 10.1016/j.fochx.2024.101981.

Macchia, M., Ceccarini, L., Molfetta, I., Cioni, P.L. and Flamini, G. (2013) Studies on saffron (*Crocus sativus* L.) from Tuscan Maremma (Italy): Effects of geographical origin, cultivation environment and drying method on volatile emission. *International Journal of Food Science & Technology* 48(11), 2370–2375. DOI: 10.1111/ijfs.12227.

Maggi, L., Carmona, M., Zalacain, A., Kanakis, C.D., Anastasaki, E. *et al.* (2010) Changes in saffron volatile profile according to its storage time. *Food Research International* 43(5), 1329–1334. DOI: 10.1016/j.foodres.2010.03.025.

Maggi, L., Carmona, M., Kelly, S.D., Marigheto, N. and Alonso, G.L. (2011) Geographical origin differentiation of saffron spice (*Crocus sativus* L. stigmas): Preliminary investigation using chemical and multi-element (H, C, N) stable isotope analysis. *Food Chemistry* 128(2), 543–548. DOI: 10.1016/j.foodchem.2011.03.063.

Maggio, A., Raimondi, G., Martino, A. and De Pascale, S. (2006) Soilless cultivation of saffron in mediterranean environment. *Acta Horticulturae 718: III International Symposium on Models for Plant Growth, Environmental Control and Farm Management in Protected Cultivation* 515–522. DOI: 10.17660/ActaHortic.2006.718.60.

Malavi, D., Nikkhah, A., Alighaleh, P., Einafshar, S., Raes, K. *et al.* (2024) Detection of saffron adulteration with *Crocus sativus* style using NIR-hyperspectral imaging and chemometrics. *Food Control* 157, 110189. DOI: 10.1016/j.foodcont.2023.110189.

Mansotra, R. and Vakhlu, J. (2022) *Crocus sativus* saffron: A 360-degree overview. In: Vakhlu, J., Ambardar, S., Alireza Salami, S. and Kole, C. (eds) *The Saffron Genome*. Springer International Publishing, Cham, Switzerland, pp. 3–25. DOI: 10.1007/978-3-031-10000-0_1.

Marinatos, N. (1985) The function and interpretation of the Theran frescoes. *Bulletin de Correspondance Hellénique. Supplément* 11(1), 219–230. DOI: 10.3406/bch.1985.5281.

Marrone, G., Urciuoli, S., Di Lauro, M., Cornali, K., Montalto, G. *et al.* (2024) Saffron (*Crocus sativus* L.) and its by-products: Healthy effects in internal medicine. *Nutrients* 16(14), 2319. DOI: 10.3390/nu16142319.

Masi, E., Taiti, C., Heimler, D., Vignolini, P., Romani, A. *et al.* (2016) PTR-TOF-MS and HPLC analysis in the characterization of saffron (*Crocus sativus* L.) from Italy and Iran. *Food Chemistry* 192, 75–81. DOI: 10.1016/j.foodchem.2015.06.090.

Mathew, B. (1982) *The Crocus: A Revision of the Genus Crocus (Iridaceae)*. B.T. Batsford, London.

Mehmeti, A., Candido, V., Canaj, K., Castronuovo, D., Perniola, M. *et al.* (2024) Energy, environmental, and economic sustainability of saffron cultivation. Insights from the first European (Italian) case study. *Sustainability* 16(3), 1179. DOI: 10.3390/su16031179.

Mir, J.I., Ahmed, N., Singh, D.B., Khan, M.H., Zaffer, S. *et al.* (2015) Breeding and biotechnological opportunities in saffron crop improvement. *African Journal of Agricultural Research* 10(9), 970–974. DOI: 10.5897/AJAR2013.8098.

Modabbernia, A., Sohrabi, H., Nasehi, A.-A., Raisi, F., Saroukhani, S. *et al.* (2012) Effect of saffron on fluoxetine-induced sexual impairment in men: Randomized double-blind placebo-controlled trial. *Psychopharmacology* 223(4), 381–388. DOI: 10.1007/s00213-012-2729-6.

Modaghegh, M.H., Shahabian, M., Esmaeili, H.A., Rajbai, O. and Hosseinzadeh, H. (2008) Safety evaluation of saffron (*Crocus sativus*) tablets in healthy volunteers. *Phytomedicine* 15(12), 1032–1037. DOI: 10.1016/j.phymed.2008.06.003.

Mohammadi, S.A. and Prasanna, B.M. (2003) Analysis of genetic diversity in crop plants—salient statistical tools and considerations. *Crop Science* 43(4), 1235–1248. DOI: 10.2135/cropsci2003.1235.

Mohammadi, H. and Reed, M. (2020) Saffron marketing: Challenges and opportunities. In: Koocheki, A. and Khajeh-Hosseini, M. (eds) *Saffron Science, Technology and Health*. Woodhead Publishing, Cambridge, UK, pp. 357–365. DOI: 10.1016/B978-0-12-818638-1.00022-8.

Mohammadpour, T., Hosseinzadeh, H., Abnous, K. and Hassanzadeh, G. (2013) Neuroprotective effect of crocin against 6-hydroxydopamine-induced cytotoxicity in SH-SY5Y cells. *Iranian Journal of Basic Medical Sciences* 16(1), 73–82.

Moise, A.R., Al-Babili, S. and Wurtzel, E.T. (2014) Mechanistic aspects of carotenoid biosynthesis. *Chemical Reviews* 114(1), 164–193. DOI: 10.1021/cr400106y.

Molina, R.V., García-Luis, A., Coll, V., Ferrer, C., Valero, M. *et al.* (2004a) Flower formation in the saffron crocus (*Crocus sativus* L.): The role of temperature. *Acta Horticulturae* 650, 39–47. DOI: 10.17660/ActaHortic.2004.650.2.

Molina, R.V., Valero, M., Navarro, Y., García-Luis, A. and Guardiola, J.L. (2004b) The effect of time of corm lifting and duration of incubation at inductive temperature on flowering in the saffron plant (*Crocus sativus* L.). *Scientia Horticulturae* 103(1), 79–91. DOI: 10.1016/j.scienta.2004.04.008.

Molina, R.V., Valero, M., Navarro, Y., Guardiola, J.L. and García-Luis, A. (2005) Temperature effects on flower formation in saffron (*Crocus sativus* L.). *Scientia Horticulturae* 103(3), 361–379. DOI: 10.1016/j.scienta.2004.06.005.

Mollafilabi, A. (2004) Experimental findings of production and echo-physiological aspects of saffron (*Crocus sativus* L.). *Acta Horticulturae 650: International Symposium on Saffron Biology and Biotechnology* 195–200. DOI: 10.17660/ActaHortic.2004.650.20.

Mollafilabi, A. (2014) *Effect of New Cropping Technologies on Growth Characteristics, Yield, Yield Components of Flower, and Corm Criteria of Saffron (Crocus sativus L.)*. University of Mashhad, Mashhad, Iran.

Mollafilabi, A. and Aslami, M.H. (2020) Replacement of saffron (*Crocus sativus* L.) with poppy (*Papaver somniferum* L.) and its socioeconomic impact in Afghanistan. In: Hirich, A., Choukr-Allah, R. and Ragab, R. (eds) *Emerging Research in Alternative Crops*. Springer International Publishing, Cham, Switzerland, pp. 359–372. DOI: 10.1007/978-3-319-90472-6_16.

Mollafilabi, A., Koocheki, A., Moeinerad, H. and Kooshki, L. (2013) Effect of plant density and weight of corm on yield and yield components of saffron (*Crocus sativus* L.) under soil, hydroponic and plastic tunnel cultivation. *Saffron Agronomy & Technology* 4, 5. DOI: 10.22048/jsat.2014.4815.

Mollafilabi, A., Davari, K. and Dehaghi, M.A. (2020) Saffron yield and quality as influenced by different irrigation methods. *Scientia Agricola* 78(1), e20190084. DOI: 10.1590/1678-992x-2019-0084.

Montanari, M. (2012) *Medieval Food Culture*. Columbia University Press, New York.

Moraga, A.R., Rambla, J.L., Ahrazem, O., Granell, A. and Gómez-Gómez, L. (2009) Metabolite and target transcript analyses during *Crocus sativus* stigma development. *Phytochemistry* 70(8), 1009–1016. DOI: 10.1016/j.phytochem.2009.04.022.

Mzabri, I., Addi, M. and Berrichi, A. (2019) Traditional and modern uses of saffron (*Crocus sativus*). *Cosmetics* 6(4), 63. DOI: 10.3390/cosmetics6040063.

Naficy, H. (2001) *An Accented Cinema: Exilic and Diasporic Filmmaking*. Princeton University Press, Princeton, NJ, pp. 114–118.

Naghizadeh, B., Boroushaki, M.T., Vahdati Mashhadian, N. and Mansouri, M.T. (2008) Protective effects of crocin against cisplatin-induced oxidative stress and nerve conduction velocity deficit in rats. *Iranian Journal of Pharmaceutical Research* 7(3), 189–196.

Negbi, M. (1990) Physiological research on the saffron crocus (*Crocus sativus*). In: Tammaro, L.F. and Marra, L. (eds) *Proceedings of the International Conference on Saffron (Crocus sativus L.)*, Universita Degli Studi, L'Aquila e Academia Italiana della Cucina, L'Aquila, Italy, pp. 183–207.

Negbi, M. (1999) Saffron cultivation: Past, present and future prospects. In: Negbi, M. (ed.) *Saffron*. CRC Press, London, pp. 1–15. DOI: 10.1201/9780203303665.

Nehvi, F.A. and Yasmin, S. (2017) Advance in saffron research for integrated development of saffron in Kashmir, India. *Acta Horticulturae* 1184(1184), 63–68. DOI: 10.17660/ActaHortic.2017.1184.9.

Nemati, Z., Blattner, F.R., Kerndorff, H., Erol, O. and Harpke, D. (2018) Phylogeny of the saffron-crocus species group, *Crocus* series *Crocus* (Iridaceae). *Molecular Phylogenetics and Evolution* 127, 891–897. DOI: 10.1016/j.ympev.2018.06.036.

Nemati, Z., Harpke, D., Gemicioglu, A., Kerndorff, H. and Blattner, F.R. (2019) Saffron (*Crocus sativus*) is an autotriploid that evolved in Attica (Greece) from wild *Crocus cartwrightianus*. *Molecular Phylogenetics and Evolution* 136, 14–20. DOI: 10.1016/j.ympev.2019.03.022.

Nisar, N., Li, L., Lu, S., Khin, N.C. and Pogson, B.J. (2015) Carotenoid metabolism in plants. *Molecular Plant* 8(1), 68–82. DOI: 10.1016/j.molp.2014.12.007.

Ochiai, T., Shimeno, H., Mishima, K., Iwasaki, K., Fujiwara, M. *et al.* (2004) Protective effects of carotenoids from saffron against neuronal injury. *Journal of Pharmacological Sciences* 95(1), 124–125.

Ordoudi, S.A. and Tsimidou, M.Z. (2004) Saffron quality: Effect of agricultural practices, processing and storage. In: Dris, R. and Jain, S.M. (eds) *Production Practices and Quality Assessment of Food Crops Volume 1: Preharvest Practice*. Springer, Dordrecht, The Netherlands, pp. 209–260. DOI: 10.1007/1-4020-2533-5_8.

Ordoudi, S.A. and Tsimidou, M. (2016) Saffron adulteration: Detection methods and market implications. *Food Chemistry* 221, 1112–1120.

Orfanou, O. and Tsimidou, M. (1996) Evaluation of the colouring strength of saffron spice by UV—vis spectrometry. *Food Chemistry* 57(3), 463–469. DOI: 10.1016/S0308-8146(96)00003-9.

Orlando, M., Trivellini, A., Puccinelli, M., Ferrante, A., Incrocci, L. *et al.* (2022) Increasing the functional quality of *Crocus sativus* L. by-product (tepals) by controlling spectral composition. *Horticulture, Environment, and Biotechnology* 63(3), 363–373. DOI: 10.1007/s13580-021-00407-1.

Orti, M.A., Gomez, L.G., Rubio, A. and Escribano, J. (2004) Development and gene expression in saffron corms. *Acta Horticulturae* 650, 35–38.

Pandita, D. (2021) Saffron (*Crocus sativus* L.): Phytochemistry, therapeutic significance and omics-based biology. In: Aftab, T. and Hakeem, K.R. (eds) *Medicinal and Aromatic Plants*. Academic Press, London, pp. 325–396.

Papandreou, M.A., Kanakis, C.D., Polissiou, M.G., Efthimiopoulos, S., Cordopatis, P. *et al.* (2006) Inhibitory activity on amyloid-β aggregation and antioxidant properties of *Crocus sativus* stigmas extract and its crocin constituents. *Journal of Agricultural and Food Chemistry* 54(23), 8762–8768. DOI: 10.1021/jf061932a.

Parizad, S., Dizadji, A., Habibi, M.K., Winter, S., Kalantari, S. *et al.* (2017) Prevalence of saffron latent virus (SaLV), a new Potyvirus species, in saffron fields of Iran. *Journal of Plant Pathology* 99, 802–802. DOI: 10.4454/JPP.V99I3.3963.

Petersen, G., Seberg, O., Thorsøe, S., Jørgensen, T. and Mathew, B. (2008) A phylogeny of the genus *Crocus* (Iridaceae) based on sequence data from five plastid regions. *Taxon* 57(2), 487–499.

Pitsikas, N. (2016) Constituents of saffron (*Crocus sativus* L.) as potential candidates for the treatment of anxiety disorders and schizophrenia. *Molecules* 21(3), 303. DOI: 10.3390/molecules21030303.

Pitsikas, N., Zisopoulou, S., Tarantilis, P.A., Kanakis, C.D., Polissiou, M.G. *et al.* (2007) Effects of the active constituents of *Crocus sativus* L., crocins on recognition and spatial rats' memory. *Behavioural Brain Research* 183(2), 141–146. DOI: 10.1016/j.bbr.2007.06.001.

Pitsikas, N., Boultadakis, A., Georgiadou, G., Tarantilis, P.A. and Sakellaridis, N. (2008) Effects of the active constituents of *Crocus sativus* L., crocins, in an animal model of anxiety. *Phytomedicine* 15(12), 1135–1139. DOI: 10.1016/j.phymed.2008.06.005.

Plessner, O., Negbi, M., Ziv, M. and Basker, D. (1989) Effects of temperature on the flowering of the saffron crocus (*Crocus sativus* L.): Induction of hysteranthy. *Israel Journal of Plant Sciences* 38(1), 1–7.

Poggi, L.M., Portela, A.J., Pontin, M.A. and Molina, R.V. (2010) Corm size and incubation effects on time to flowering and threads yield and quality in saffron production in Argentina. *Acta Horticulturae 850: III International Symposium on Saffron: Forthcoming Challenges in Cultivation, Research and Economics* 193–198. DOI: 10.17660/ActaHortic.2010.850.32.

Rahimi, H. and Kamali, K. (1993) Laboratory studies on biology of bulb mite *Rhizoglyphus robini* (Acari: Acaridae) and its damages on saffron corm in Gonabad and Qaen. *Plant Protection (Scientific Journal of Agriculture)* 16(1 & 2), 53–63.

Rahimi, H., Nateq Golestan, M. and Kakhki, A. (2018) Exploring the impact of non-chemical methods on the control of bulb mite (*Rhizoglyphus robini* Claparáde) and yield of saffron (*Crocus sativus* L.). *New Zealand Plant Protection* 71, 93–101. DOI: 10.30843/nzpp.2018.71.108.

Raina, A., Kaul, S. and Dhar, M.K. (2024) Sniffing out adulteration in saffron: Detection methods and health risks. *Food Control* 155, 110042. DOI: 10.1016/j.foodcont.2023.110042.

Rastgoo, M., Hosseinzadeh, H., Alavizadeh, H., Abbasi, A., Ayati, Z. *et al.* (2014) Crocin, a constituent of saffron, improves memory and learning in a rat model of Alzheimer's disease. *Journal of Pharmacopuncture* 17(3), 50–56.

Rees, A.R. (1992) *Ornamental Bulbs, Corms and Tubers*. Butterworth-Heinemann, London. DOI: 10.1079/9780851986562.0000.

Rezvani-Moghaddam, P. (2020) Ecophysiology of saffron. In: Koocheki, A. and Khajeh-Hosseini, M. (eds) *Saffron Science, Technology and Health*. Woodhead Publishing, Cambridge, UK, pp. 119–137.

Rigi, H., Mohtashami, L., Asnaashari, M., Emami, S.A. and Tayarani-Najaran, Z. (2021) Dermoprotective effects of saffron: A mini review. *Current Pharmaceutical Design* 27(46), 4693–4698. DOI: 10.2174/1381612827666210920150855.

Rubio-Moraga, A., Castillo-López, R., Gómez-Gómez, L. and Ahrazem, O. (2009) Saffron is a monomorphic species as revealed by RAPD, ISSR and microsatellite analyses. *BMC Research Notes* 2, 1–5. DOI: 10.1186/1756-0500-2-189.

Rukšāns, J. (2017) *The World of Crocuses*. The Latvian Academy of Sciences, Riga.

Rukšāns, J. (2023) *The World of Crocuses: The First Supplement*. Latvian Academy of Sciences, Riga.

Russo, M., Castellano, M.A. and Martelli, G.P. (1979) Rhabdovirus-like particles in English ivy (*Hedera helix*) and ivy-leafed geranium (*Pelargonium peltatum*). *Journal of Phytopathology* 96(2), 122–131. DOI: 10.1111/j.1439-0434.1979.tb01630.x.

Ruttig, T. (2014) Herat's saffron boom: An alternative to opium? Afghanistan analysts Network.

Sahakian, S. (ed.) (2015) *Crocologia – A Detailed Study of Saffron, the King of Plants*. Skylight Press, London.

Salami, S.A. and Husaini, A.M. (2022) Genetic mapping and molecular markers in saffron. In: Vakhlu, J., Ambardar, S., Alireza Salami, S. and Kole, C. (eds) *The Saffron Genome*. Springer International Publishing, Cham, Switzerland, pp. 83–94.

Salas, M. del C., Montero, J.L., Diaz, J.G., Berti, F., Quintero, M.F. *et al.* (2020) Defining optimal strength of the nutrient solution for soilless cultivation of saffron in the Mediterranean. *Agronomy* 10(9), 1311. DOI: 10.3390/agronomy10091311.

Salehi, A., Shariatifar, N., Pirhadi, M. and Zeinali, T. (2022) An overview on different detection methods of saffron (*Crocus sativus* L.) adulterants. *Journal of Food Measurement and Characterization* 16(6), 4996–5006. DOI: 10.1007/s11694-022-01586-w.

Samarghandian, S., Azimi-Nezhad, M. and Farkhondeh, T. (2017) Immunomodulatory and antioxidant effects of saffron aqueous extract (*Crocus sativus* L.) on streptozotocin-induced diabetes in rats. *Indian Heart Journal* 69(2), 151–159. DOI: 10.1016/j.ihj.2016.09.008.

Sánchez, A.M., Carmona, M., Jarén-Galán, M., Mosquera, M.I.M. and Alonso, G.L. (2011) Picrocrocin kinetics in aqueous saffron spice extracts (*Crocus sativus* L.) upon thermal treatment. *Journal of Agricultural and Food Chemistry* 59(1), 249–255. DOI: 10.1021/jf102828v.

Sanjuán-López, A.I. and Resano Ezcaray, H. (2020) Labels for a local food speciality product: The case of saffron. *Journal of Agricultural Economics* 71(3), 778–797. DOI: 10.1111/1477-9552.12376.

Saxena, R.B. (2010) Botany, taxonomy and cytology of *Crocus sativus* series. *AYU (An International Quarterly Journal of Research in Ayurveda)* 31(3), 374. DOI: 10.4103/0974-8520.77153.

Schmidt, M., Betti, G. and Hensel, A. (2007) Saffron in phytotherapy: Pharmacology and clinical uses. *Wiener Medizinische Wochenschrift (1946)* 157(13–14), 315–319. DOI: 10.1007/s10354-007-0428-4.

Schmidt, T., Heitkam, T., Liedtke, S., Schubert, V. and Menzel, G. (2019) Adding color to a century-old enigma: Multi-color chromosome identification unravels the autotriploid nature of saffron (*Crocus sativus*) as a hybrid of wild *Crocus cartwrightianus* cytotypes. *The New Phytologist* 222(4), 1965–1980. DOI: 10.1111/nph.15715.

Sereshti, H., Poursorkh, Z., Aliakbarzadeh, G. and Zarre, S. (2018) Quality control of saffron and evaluation of potential adulteration by means of thin layer chromatography-image analysis and chemometrics methods. *Food Control* 90, 48–57. DOI: 10.1016/j.foodcont.2018.02.026.

Shafiee, A., Arekhi, S., Omranzadeh, A. and Sahebkar, A. (2016) Saffron in the treatment of depression, anxiety and other mental disorders: A review. *Journal of Psychopharmacology* 30(2), 102–113.

Shafiee, A., Jafarabady, K., Seighali, N., Mohammadi, I., Rajai Firouz Abadi, S. *et al.* (2025) Effect of saffron versus selective serotonin reuptake inhibitors (SSRIs) in treatment of depression and anxiety: A meta-analysis of randomized controlled trials. *Nutrition Reviews* 83(3), e751–e761. DOI: 10.1093/nutrit/nuae076.

Shahrokhi, M.B., Rahimi, H. and Rashed, M.H. (2006) Saffron pests, diseases, and weeds. In: Kafi, M. (ed.) *Saffron (Crocus Sativus): Production and Processing*. CRC Press, Boca Raton, FL, pp. 91–109.

Shamsa, A., Hosseinzadeh, H., Molaei, M., Shakeri, M.T. and Rajabi, O. (2009) Evaluation of *Crocus sativus* L. (saffron) on male erectile dysfunction: A pilot study. *Phytomedicine* 16(8), 690–693. DOI: 10.1016/j.phymed.2009.03.008.

Shamshiri, N., Fattahi, R., Mani-Varnosfaderani, A., Barzegar, M. and Sahari, M.A. (2024) Geographical authentication of saffron by chemometrics applied to the ion mobility spectrometry data. *Food Chemistry: X* 22, 101455. DOI: 10.1016/j.fochx.2024.101455.

Sharma, M., Kaul, S. and Dhar, M.K. (2019) Transcript profiling of carotenoid/apocarotenoid biosynthesis genes during corm development of saffron (*Crocus sativus* L.). *Protoplasma* 256(1), 249–260. DOI: 10.1007/s00709-018-1296-z.

Sheng, L., He, S., Qian, Z. and Zheng, S. (2006) Protective effects of crocin on acute cerebral ischemia-reperfusion injury in rats. *Chinese Pharmacological Bulletin* 22(4), 440–443.

Shokrpour, M. (2019) Saffron (*Crocus sativus* L.) breeding: Opportunities and challenges. In: Al-Khayri, J., Jain, S. and Johnson, D. (eds) *Advances in Plant Breeding Strategies: Industrial and Food Crops*. Springer, Cham, Switzerland, pp. 675–706. DOI: 10.1007/978-3-030-23265-8_17.

Shokrpour, M., Abedi, Z., Kalantari, S. and Salami, S.A. (2017) Study of genetic variation in some Iranian saffron accessions using molecular markers of RAPD and ISSR. *Saffron Agronomy and Technology* 4(4), e257–Pe265.

Si, W., Xiong, L., Zhou, H., Wu, H., Liu, Z. *et al.* (2022) Comprehensive characterization of ingredients in *Crocus sativus* L. from different origins based on the combination of targeted and nontargeted strategies. *Food Chemistry* 397, 133777. DOI: 10.1016/j.foodchem.2022.133777.

Siddique, H.R., Fatma, H. and Khan, M.A. (2020) Medicinal properties of saffron with special reference to cancer: A review of preclinical studies. In: Sarwat, M. and Sumaiya, S. (eds) *Saffron: The Age-Old Panacea in a New Light*. Academic Press, pp. 233–244.

Siracusa, L., Gresta, F., Avola, G., Albertini, E., Raggi, L. *et al.* (2013) Agronomic, chemical and genetic variability of saffron (*Crocus sativus* L.) of different origin by LC-UV–vis-DAD and AFLP analyses. *Genetic Resources and Crop Evolution* 60(2), 711–721. DOI: 10.1007/s10722-012-9868-9.

Soeda, S., Ochiai, T., Paopong, L., Tanaka, H., Shoyama, Y. *et al.* (2001) Crocin suppresses tumor necrosis factor-alpha-induced cell death of neuronally differentiated PC-12 cells. *Life Sciences* 69(24), 2887–2898. DOI: 10.1016/S0024-3205(01)01357-1.

Song, W., Wei, X., Wang, H., Xu, J., Tang, X. *et al.* (2024) Rapid and low-cost detection of saffron (*Crocus sativus* L.) adulteration using smartphone videos and spectral data fusion strategy. *Journal of Food Composition and Analysis* 135, 106691. DOI: 10.1016/j.jfca.2024.106691.

Soorni, A., Fatahi, R., Haak, D.C., Salami, S.A. and Bombarely, A. (2017) Assessment of genetic diversity and population structure in Iranian cannabis germplasm. *Scientific Reports* 7(1), 15668. DOI: 10.1038/s41598-017-15816-5.

Spence, C. (2023) Saffron: The colourful spice. *International Journal of Gastronomy and Food Science* 34, 100821. DOI: 10.1016/j.ijgfs.2023.100821.

Srivastava, R., Ahmed, H., Dixit, R.K. and Dharamveer and Saraf, S.A. (2010) *Crocus sativus* L.: A comprehensive review. *Pharmacognosy Reviews* 4(8), 200–208. DOI: 10.4103/0973-7847.70919.

Sud, A.K., Paul, Y.S. and Thakur, B.R. (1999) Corm rot of saffron and its management. *Journal of Mycology and Plant Pathology* 29(3), 380–382.

Tamaddonfard, E. and Hamzeh-Gooshchi, N. (2010) Effect of crocin on the morphine-induced antinociception in the formalin test in rats. *Phytotherapy Research* 24(3), 410–413. DOI: 10.1002/ptr.2965.

Tammaro, F. (1999) Saffron (*Crocus sativus* L.) in Italy. In: Negbi, M. (ed.) *Saffron*. CRC Press, London, pp. 48–55.

Terro, C., Strojnik, L., Potočnik, D., Hudobivnik, M.J., Mazej, D. *et al.* (2025) Stable isotope and multi-elemental analysis combined with statistical modeling for saffron origin authentication. *Food Chemistry X* 103196. DOI: 10.1016/j.fochx.2025.103196.

Tosan, M., Khashei Siuki, A., Sangari, M. and Rezvani Moghaddam, P. (2024) Analysis of the global research trend of saffron (*Crocus sativus* L.) between 2000–2023. *Saffron Agronomy and Technology* 12(2), 115–138. DOI: 10.22048/JSAT.2024.443037.1524.

Touliatos, D., Dodd, I.C. and McAinsh, M. (2016) Vertical farming increases lettuce yield per unit area compared to conventional horizontal hydroponics. *Food and Energy Security* 5(3), 184–191. DOI: 10.1002/fes3.83.

Toussaint-Samat, M. (2008) *A History of Food*, 2nd edn. Blackwell, Oxford, UK, pp. 512–516. DOI: 10.1002/9781444305135.

Tsaftaris, A., Pasentsis, K., Makris, A., Darzentas, N., Polidoros, A. *et al.* (2011) The study of the E-class SEPALLATA3-like MADS-box genes in wild-type and mutant flowers of cultivated saffron crocus (*Crocus sativus* L.) and its putative progenitors. *Journal of Plant Physiology* 168(14), 1675–1684. DOI: 10.1016/j.jplph.2011.03.015.

Ünaldı, E.Ü. (2007) A plant exposed to danger: Saffron (*Crocus sativus* L.). *Fırat Üniversitesi Sosyal Bilimler Dergisi* 17(2), 53–69.

United Nations Industrial Development Organization (UNIDO) (2014) Saffron Industry Value Chain Development in Iran: Diagnostic Study Report. Saffron Industry Value Chain Development. SAP ID 120595. Available at: https://www.unido.org (accessed 20 January 2026).

United Nations Office on Drugs and Crime (UNODC) (2021) Afghanistan Opium Survey: Cultivation and Production. Prepared by the Research and Trend Analysis Branch, United

Nations Office on Drugs and Crime. Available at: https://www.unodc.org/documents/crop-monitoring/Afghanistan/Afghanistan_Opium_Survey_2021.pdf (accessed 23 April 2026).

Vahedi, M., Kabiri, M., Salami, S.A., Rezadoost, H., Mirzaie, M. *et al.* (2018) Quantitative HPLC-based metabolomics of some Iranian saffron (*Crocus sativus* L.) accessions. *Industrial Crops and Products* 118, 26–29. DOI: 10.1016/j.indcrop.2018.03.024.

Vahedi, M., Salami, S.A., Shokrpour, M. and Rezadoost, H. (2019) Comparative performance of transcriptome assembly programs for saffron (*Crocus sativus* L.). *Saffron Agronomy and Technology* 7(1), e69–Pe80.

Valle García-Rodríguez, M., López-Córcoles, H. and Alonso, G.L. (2017a) Effect of the hydroponic growing of forced *Crocus sativus* L. on the saffron quality. In: Lage, M., Al Faiz, C. and Badraoui, M. (eds) *V International Symposium on Saffron Biology and Technology: Advances in Biology, Technologies, Uses and Market*, pp. 287–292 (Vol. 1184). DOI: 10.17660/ActaHortic.2017.1184.41.

Valle García-Rodríguez, M., López-Córcoles, H., Alonso, G.L., Pappas, C.S., Polissiou, M.G. *et al.* (2017b) Comparative evaluation of an ISO 3632 method and an HPLC-DAD method for safranal quantity determination in saffron. *Food Chemistry* 221, 838–843. DOI: 10.1016/j.foodchem.2016.11.089.

Valle García-Rodríguez, M., Serrano-Díaz, J., Tarantilis, P.A., López-Córcoles, H., Carmona, M. *et al.* (2014) Determination of saffron quality by high-performance liquid chromatography (HPLC-DAD). *Journal of Agricultural and Food Chemistry* 62(32), 8068–8074. DOI: 10.1021/jf5019356.

Varliklioz Er, S., Eksi-Kocak, H., Yetim, H. and Boyaci, I.H. (2017) Novel spectroscopic method for determination and quantification of saffron adulteration. *Food Analytical Methods* 10(5), 1547–1555. DOI: 10.1007/s12161-016-0710-4.

Villa, C., Costa, J., Oliveira, M.B.P.P. and Mafra, I. (2017) Novel quantitative real-time PCR approach to determine safflower (*Carthamus tinctorius*) adulteration in saffron (*Crocus sativus*). *Food Chemistry* 229, 680–687. DOI: 10.1016/j.foodchem.2017.02.136.

Visit Saffron Walden (2022) Medieval Walden [Visitor information leaflet]. Visit Saffron Walden. Available at: https://www.visitsaffronwalden.gov.uk/2022/01/medieval-walden-2/ (accessed 29 April 2026).

Wang, C.J., Hwang, L.S., Lin, J.K. (1984) Reversible hepatic black pigmentation and enzyme alteration induced by prolonged feeding of high dose of crocin dyes in rats. *Proc Natl Sci Counc Repub China B.* 8(3), 246-253. PMID: 6571591.

Walter, M.H. and Strack, D. (2011) Carotenoids and their cleavage products: biosynthesis and functions. *Natural Product Reports* 28(4), 663. DOI: 10.1039/c0np00036a.

Wang, Z., Li, X., Xu, J., Yang, Z. and Zhang, Y. (2021) Effects of ambient temperature on flower initiation and flowering in saffron (crocus sativus l.). *Scientia Horticulturae* 279, 109859. DOI: 10.1016/j.scienta.2020.109859.

Warren, P. (2005) Flowers for the goddess? New fragments of wall paintings from Knossos. *British School at Athens Studies* 13, 131–148.

Wikimedia Commons (2026) Saffron Gatherers (Akrotiri). Available at: https://commons.wikimedia.org/wiki/Category:Saffron_gatherers_(Akrotiri) (accessed 23 April 2026).

Willard, P. (2001) *Secrets of Saffron: The Vagabond Life of the World's Most Seductive Spice*. Beacon Press, Boston, MA.

Winterhalter, P. and Straubinger, M. (2000) Saffron—renewed interest in an ancient spice. *Food Reviews International* 16(1), 39–59. DOI: 10.1081/FRI-100100281.

World Bank (2011) *Understanding Gender in Agricultural Value Chains: The Cases of Grapes/Raisins, Almonds and Saffron in Afghanistan*. The World Bank, Washington, DC.

Wright, C.W. (1999) *A Mediterranean Feast*. William Morrow, New York, pp. 62–65. DOI: 10.1016/S0307-4412(98)00170-8.

Xie, L., Luo, Z., Jia, X., Mo, C., Huang, X. *et al.* (2023) Synthesis of crocin I and crocin II by multigene stacking in *Nicotiana benthamiana*. *International Journal of Molecular Sciences* 24(18), 14139. DOI: 10.3390/ijms241814139.

Xu, G., Gong, Z., Yu, W., Gao, L., He, S. *et al.* (2006) The protective effects of crocetin on acute cerebral ischemia and reperfusion injury in rats. *Chinese Pharmacological Bulletin* 22(11), 1365–1368.

Xu, G., Gong, Z., Yu, W., Gao, L., He, S. *et al.* (2009) Increased expression of aquaporin 4 in the rat brain after transient focal cerebral ischemia and reperfusion: Effect of crocetin. *Journal of Molecular Neuroscience* 39(1–2), 48–55.

Xu, Z., Chen, S., Wang, Y., Tian, Y., Wang, X. *et al.* (2024) *Crocus* genome reveals the evolutionary origin of crocin biosynthesis. *Acta Pharmaceutica Sinica B* 14(4), 1878–1891. DOI: 10.1016/j.apsb.2023.12.013.

Yarami, N., Kamgar-Haghighi, A.A., Sepaskhah, A.R. and Zand-Parsa, S. (2011) Determination of the potential evapotranspiration and crop coefficient for saffron using a water-balance lysimeter. *Archives of Agronomy and Soil Science* 57(7), 727–740. DOI: 10.1080/03650340.2010.485985.

Yasmin, S. and Nehvi, F.A. (2013) Saffron as a valuable spice: A comprehensive review. *African Journal of Agricultural Research* 8(3), 234–242.

Yüzbaşioğlu, S. (2019) *Crocus keltepensis* (Iridaceae): A new *Crocus* species from the highest part of the Samanlı Mountains, NW Turkey. *Phytotaxa* 418(2), 230–236. DOI: 10.11646/phytotaxa.418.2.8.

Zaffar, G., Wani, S.A., Anjum, T. and Zeerak, N.A. (2004) Colchicine induced variability in saffron. *Acta Horticulturae 650: I International Symposium on Saffron Biology and Biotechnology* 277–280. DOI: 10.17660/ActaHortic.2004.650.31.

Zaffar, G., Wani, S.A. and Shikari, A.B. (2008) Corm development and flower formation in saffron. In: *Saffron Production in Jammu and Kashmir*. Directorate of Extension Education SKUAST-K, Srinagar, Jammu and Kashmir, India, pp. 87–98.

Zheng, Y.Q., Liu, J.X., Wang, J.N. and Xu, L. (2007) Effects of crocin on reperfusion-induced oxidative/nitrative injury to cerebral microvessels after global cerebral ischemia. *Brain Research* 1138, 86–94. DOI: 10.1016/j.brainres.2006.12.064.

Printed and bound by CPI Group (UK) Ltd, Croydon, CR0 4YY
03/08/2026
14934317-0001